Secondary Recovery and Carbonate Reservoirs

Secondary Recovery and Carbonate Reservoirs

Gerald L. Langnes
Mobil Oil Corporation
New York, New York

John O. Robertson, Jr.
University of Southern California
Los Angeles, California

George V. Chilingar
University of Southern California
Los Angeles, California

Foreword by NICK VAN WINGEN

**American Elsevier
Publishing Company, Inc.**
NEW YORK

AMERICAN ELSEVIER PUBLISHING COMPANY, INC.
52 Vanderbilt Avenue, New York, N.Y. 10017

ELSEVIER PUBLISHING COMPANY
335 Jan Van Galenstraat, P.O. Box 211
Amsterdam, The Netherlands

International Standard Book Number 0-444-00116-6

Library of Congress Card Number 70-168265

Copyright © 1972 by American Elsevier Publishing Company, Inc.

All rights reserved.
No part of this publication may be reproduced,
stored in a retrieval system, or transmitted
in any form or by any means, electronic,
mechanical, photocopying, recording,
or otherwise, without the prior
written permission of the publisher,
American Elsevier Publishing Company, Inc.,
52 Vanderbilt Avenue, New York, New York 10017.

Manufactured in the United States of America

*We dedicate this book to our
patient and understanding wives*
MARY M. LANGNES
KAREN M. ROBERTSON
YELBA M. CHILINGAR

and

to the outstanding scientists
ACADEMICIAN G. I. TEODOROVICH
PROFESSOR SYLVAIN J. PIRSON

Contents

Foreword by Nick van Wingen................................. xi

Acknowledgments... xiii

Chapter 1

Introduction to the Technology of Secondary Recovery

Introduction...	1
Goals of the Developers and the Political-Economic Environment........	3
Characteristics of the Recovery Processes and the Reservoirs............	5
Application Ranges for Secondary Recovery Processes.................	16
Questions and Problems..	17
References...	18

Chapter 2

Planning a Secondary Recovery Project

Introduction...	20
Collecting Data for Planning a Secondary Recovery Project............	20
Laboratory Investigation..	27
Data Processing by Computer...................................	38
Pilot Tests and their Value.....................................	40
Questions and Problems..	42
References...	43

Chapter 3

General Principles of Waterflood Design

Introduction...	45
Waterflooding "Rules of Thumb".................................	45
Flood Design by Analogy.......................................	51
Sweep Efficiency..	55
Major Predictive Techniques....................................	74
Improved Waterflood Processes..................................	90
Questions and Problems..	94
References...	95

Chapter 4

Carbonate Reservoir Waterflood Predictions and Performance

Introduction	99
Intercrystalline-Intergranular Porosity Systems	99
Fracture-Matrix Porosity Systems	113
Vugular-Solution Porosity Systems	122
Questions and Problems	126
References	127

Chapter 5

Gas Injection—Immiscible Displacement

Introduction	130
Predictive Techniques	131
Reservoir Performance	138
Questions and Problems	143
References	143

Chapter 6

Miscible Flooding

Introduction	145
Miscibility	145
Sweep Efficiency	147
High-Pressure Gas Injection	148
Enriched-Gas Drive	150
Liquid Petroleum Gas (LPG) Slug Drive	155
Predictive Techniques	158
Intergranular-Intercrystalline Porosity System	159
Fracture-Matrix Porosity System	162
Vugular-Solution Porosity System	162
Questions and Problems	163
References	164

Appendix A

Determination of Volumes and Average Heights of Fractures

Introduction	166
Derivation of Tank Oil-In-Place in Fractured Reservoirs	167
Determination of Average Height of Fractures	171
Sample Problem	172
References	172

Contents

Appendix B
Fundamentals of Surface and Capillary Forces

Introduction	173
Interfacial Tension and Contact Angle	173
Effect of Contact Angle and Interfacial Tension on Movement of Oil	178
References	182

Appendix C
Formation Volume Factors for Natural Gas, Water, and Crude Oils

Formation Volume Factor for Natural Gas	183
Formation Volume Factor for Water	185
Formation Volume Factor for Crude Oils	185
References	195

Appendix D
Viscosities of Air, Water, Natural Gas, and Crude Oil

Introduction	197
Viscosity of Air	197
Viscosity of Water	197
Viscosity of Natural Gas	197
Viscosity of Crude Oil	206
References	206

Appendix E
Rock and Fluid Compressibilities

Introduction to Estimating Reservoir Compressibilities	207
Gas Reservoirs	212
Oil Reservoirs above the Bubble Point	213
Oil Reservoirs below the Bubble Point	213
Oil Reservoirs where Gas Saturation is Less Than Critical	214
Calculation of Compressibilities for the Various Phases	214
References	223

Appendix F
Relative Permeability Concepts

Introduction	225
Effect of Polarity of Oil and Water Hardness on Relative Permeability Curves	227
Effect of Water Chemistry on Recovery	230
Effect of Carbonate Material in Porous Medium on Relative Permeabilities	231
Effect of Temperature on Relative Permeability	232
References	242

Appendix G

Relationships Among Surface Area, Permeability and Porosity

Introduction	243
Derivation of Theoretical Equation Relating Porosity, Permeability, and Surface Area	243
Statistical Technique of Determining Specific Surface Area	249
Relationships among Rock Granulometric Composition, Porosity, and Permeability	251
References	252
Bibliography	252

Appendix H

Carbonate Reservoir Data for Text Problems	253
References	263

Appendix I

Conversion of Units

Theoretical Aspects	264
Example 1. Dynamic Viscosity Conversion Factor	264
Example 2. Determination of Multicomponent Conversion Constants	266
Temperature Conversion Formulas	286
Bibliography	287

Author Index	289
Subject Index	295

Foreword

The future availability of crude petroleum products will be progressively more dependent upon the petroleum industry's ability to improve the recovery efficiency for existing oil accumulations by the application of secondary recovery techniques. This situation is the result of the inability to discover adequate amounts of economic new reserves to replace the vast amounts of oil used annually.

The objective of this book is to present methods for the prediction and interpretation of the detailed behavior of oil reservoirs under the different known methods of secondary recovery. The subject is complex, as the fundamental mechanism of flow within the reservoir rock material is controlled by potential gradients and gravity and interfacial forces, acting on a three-phase system (oil, water, and gas).

Much literature has appeared on the subject of secondary recovery, but only in a few instances have the published data been summarized in book form. Furthermore, only a small portion of the information deals specifically with carbonate reservoirs. This work, therefore, fills an important gap by combining in one reference volume all of the available knowledge which pertains to this subject. The importance of the study is apparent when it is considered that approximately 45% of the world's production of petroleum is derived from carbonate formations. The book has been designed to serve both as a textbook for students and as a handbook for practicing engineers.

It is important to note that the comprehensive Appendix prepared by the writers will be of great value because it covers some of the fundamental aspects of reservoir engineering indispensable for any serious study of secondary recovery processes.

NICK VAN WINGEN
Consulting Petroleum Engineer
South Pasadena, California

*The authors would like
to extend their appreciation to*

Miss Shirley Nickelson
    the typist

Mr. Cliff Mathieson
    the draftsman

and to

Professor Lyman L. Handy
    for his encouragement

CHAPTER 1

Introduction to the Technology of Secondary Recovery

Introduction

The American Petroleum Institute (API) glossary defines primary recovery as the oil, gas, or oil and gas recovered by any method (natural flow or artificial lift) that may be used to produce them through a single wellbore; the fluid enters the wellbore by action of the reservoir's native energy. The definition for secondary recovery is similar to that of primary except that more than one wellbore may be involved, and the reservoir's native energy is augmented by the injection of fluids with or without heat. The petroleum industry literature also shows a more restricted definition of secondary recovery. The term "secondary recovery" is used only when injection is started late in the life of the field, i.e., when the primary energy is nearly exhausted. The term "pressure maintenance" is used when the injection is started early in the life of the field. Inasmuch as clearcut guidelines do not exist as to where the break between pressure maintenance and secondary recovery occurs, the broader definition of secondary recovery, presented initially, is used in this book.

The goal of this book is to develop an understanding of the chacteristics of the secondary recovery processes and their application. This understanding is dependent upon the knowledge of the processes, the characteristics of the reservoirs, the political-economic environment in which the reservoirs exist, and the goals of the developers of the reservoirs. This book is primarily concerned with the characteristics of the processes and the reservoirs. The political-economic environment and the goals of the developers are treated in broad generalities. Detailed economic analyses of the processes are not presented, as they cannot be made sufficiently general to be of use across the broad spectrum of environments in which the carbonate reservoirs exist.

General and overall observations on the processes and their application are presented in the first two chapters. Figures 1-1 and 1-2, can be used as screening devices in the determination of the suitability of processes for the particular reservoir in question. The second chapter presents the steps to consider in the design of a secondary recovery project. The last four chapters cover in greater detail the processes and their specific applications to carbonate reservoirs. The discussions of the reservoirs are keyed to the various porosity systems: (1) intergranular–intercrystalline; (2) fracture–matrix; and (3) vugular–solution. The Appendixes present additional material on specific reservoir rock and fluid characteristics.

Carbonate pools with intergranular–intercrystalline porosity are comparable to unfractured sandstone reservoirs in behavior. Oolites, broken fossils, and other transported calcareous particles make up the basic building blocks. The sedimentary units may also be crystals precipitated from solution. Dolomitization may give rise to interrhombohedral porosity.

Reservoirs with fracture–matrix porosity have a well-developed double-porosity system. The matrix is usually of low permeability and contains most of the oil. The fractures contribute a minor portion of the total hydrocarbon pore space, but allow the reservoir to produce at economic rates. Strong directional permeability may also exist. Pool performance is markedly different than in the case of sandstone reservoirs. Additional data on fracture porosity are presented in Appendix A.

Vugular–solution porosity systems are more complex than either of the previous two porosity systems discussed. The original pore structure has usually been altered by the formation of solution cavities. Fractures may exist, but they do not dominate the mechanics of flow. Reefs are also in-

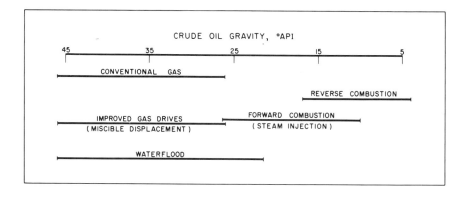

FIG. 1-1. Crude oil gravity application ranges for various secondary recovery processes. (After Roberts and Walker,[2] courtesy of the World Petroleum Congress.) °API = (141.5/sp.gr. at 60°F)−131.5.

Introduction to the Technology of Secondary Recovery

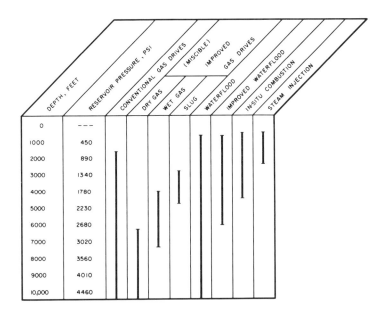

FIG. 1-2. Depth and pressure application ranges for various secondary recovery processes.

cluded in this category. Pool performance may closely resemble sandstone pool performance, but it usually demonstrates wide divergence.

There should be a single development and operating plan for each reservoir that will optimize its profit potential. This plan cannot be formulated, however, until sufficient knowledge is developed concerning the reservoir system to identify its parts and otherwise describe it.[1] The following should be thoroughly evaluated: (1) the characteristics of the recovery processes and the reservoirs, (2) the political-economic environment, and (3) the goals of the developers.

Goals of the Developers and the Political-Economic Environment

The decision as to whether to use water, gas, or a miscible type process is not entirely based upon reservoir characteristics because it is frequently governed by economics and local conservation laws. For example, in an area where large volumes of gas are produced with either no sales outlet or where the value of gas is low, the injection of gas might be the most economical method to improve oil recovery even though water injection may yield a higher recovery. If there are strong local conservation laws, stating that unsalable gas cannot be released into the atmosphere, then gas injection or possibly miscible flooding is mandatory.

If two or more methods to improve recovery appear to be applicable to the reservoir, the final selection of the process will depend on the economic evaluation of the methods being considered.[2]

The production practices used during the life of a reservoir and a portion of the political-economic environment evolve from the goals of the developers, the royalty interest owners, and the working interest owners. If the royalty interest is held by one or more individuals, maximization of current income may be their primary goal. If the reservoir is located under land which is becoming valuable for surface uses, their goal may be to finish operations as soon as possible. The preceding goals generally influence an operator to consider the early application of a secondary recovery process.

The sliding scale royalty interest system (royalty payment dependent upon production rate), such as used in California, evolved as a result of the merger of long- and short-range goals. The short-range goal is maximum current income, whereas the long-range goal is maximum return from the reservoir. The sliding scale royalty is an attempt to make both goals compatible by tying the royalty to an "ability to pay" concept, i.e., when the wells are flowing at high rates, the working interest owners can afford to pay more than when the wells are producing at low rates under artificial lift. The implementation of a secondary recovery project may be restricted and profits delayed if there is no provision for the reduction of royalty in the case of secondary recovery projects.

Kaufman[3] points out that governmental leasing practices are designed to promote optimum resource development consistent with the goals of the country. Some countries are interested in maximum governmental revenue, others desire low cost of hydrocarbons for their industries' needs, and still others are interested in the industrial development associated with the development of their natural resources.

In most countries, the leasing practices are the same for offshore and onshore tracts, the major exceptions being within the United States. Tables 1-1 through 1-3 present summaries of leasing regulations for various countries. Competitive bidding, which is used in the United States and various other areas for lease disposal, has given the major economic advantage to the government of substituting market forces for administrative judgment.[3] The major governmental advantage of a negotiated concession system is the flexibility in accomplishing other national objectives. A lower bonus may be traded for a commitment to some regional development which will be mutually beneficial to the company and the country.

The working interest owner may have many different goals. He may be concerned with return on the capital invested and set rate of return goals for individual projects, as well as for the corporation as a whole. Fluid injection projects having a low rate of return may be eliminated by this goal. If, however, crude self-sufficiency and the profit-investment ratio are of greater importance to the company, secondary recovery projects with low rate of return may be acceptable. When developers find themselves in a poor position for long-range crude supplies, they could benefit by accepting a low rate of return process which yields a marked increase in recovery.

The capacity and price elasticity of the market are two major areas of concern in the political-economic environment. If a market is already saturated with crude oil and an increase in supply will result in price decrease, little incentive may exist for starting fluid injection projects, which lead to rapid increases in production rate. Proration policies are also of major importance when considering secondary recovery. Most agencies offer incentives to initiate secondary recovery projects for the purpose of increasing the ultimate recovery of oil reserves and allow the operator to recover the additional installation and operating costs, but not necessarily to give the operator a faster payout on his capital investment. Conservation regulations are generally designed to increase the total (cumulative) production from the pool; the current rate may or may not be increased. The proration system in Alberta, Canada, however, promotes the early application of the secondary recovery process which will yield the highest ultimate recovery, because the allowable production rate is a function of the reservoir reserves. Proration systems which do not recognize that some processes cost more to apply than others and, therefore, do not offer additional incentives, promote the delay in the application of the more expensive processes and result in a lower ultimate recovery.

Characteristics of the Recovery Processes and the Reservoirs

The effectiveness of any secondary process is dependent on three major factors:

1. Saturation: the amount of oil contained within the reservoir when the process is initiated.

TABLE 1-1

Summary of Offshore Regulations in the United States[a]

Regulation	Federal	Alaska	California	Louisiana	Texas
Disposal system	Competitive	Competitive	Competitive	Competitive	Competitive
Size of lease	5760 acres	5760 acres	5760 acres	5000 acres	Variable
Relinquishment	None	None; holdings may not exceed 500,000 acres.	None	None	None
Term of lease	5 yr or as long as producing	10 yr or as long as producing	20 yr or as long as producing	3 yr or as long as producing	Variable; usually 2–5 yr.
Bonus	Variable	Variable	Variable	Variable	Variable
Royalty	Usually $16\tfrac{2}{3}\%$	5% on discovery lease for 10 yr, $12\tfrac{1}{2}\%$ thereafter and on all other leases.	Sliding scale from $16\tfrac{2}{3}$–50% on oil, depending on production; $16\tfrac{2}{3}\%$ on gas.	$12\tfrac{1}{2}\%$ minimum	$16\tfrac{2}{3}\%$ usually

Rental	Usually $3/acre annually	$1/acre until discovery	None	½ of the cash bonus offered	10¢/acre 1st yr, 25¢/acre 2nd yr, 50¢/acre 3rd yr, $1/acre 4th yr and until production.
Income tax	48% of net income offset by a depletion allowance of 22% of gross income (limited by 50% of net).	Not available	Not available	Not available	Not available
Exploration commitment	None	None	None, but lease provides for a 3-yr drilling term.	None	None
Government participation	None	None	None	None	None

[a] After Kaufman,[3] courtesy of the SPE of AIME.

TABLE 1-2
Summary of Offshore Regulations and Terms of Contracts or Agreements in the Middle East[a]

Regulation	Abu Dhabi	Iran	Kuwait[b]	Neutral Zone	Saudi Arabia	U.A.R.
Disposal system	Negotiated	Negotiated	Negotiated	Negotiated	Negotiated	Negotiated
Size of lease	Variable	Variable	Variable	Variable	Variable	Variable
Relinquishment	25% within 5 yr, 25% next 3 yr, 25% next 2 yr.	25% by 5th yr, 25% next 5 yr, all nonproducing acreage by 12th yr.	20% by 3rd yr, 20% every 5 yr, thereafter until all nonproducing acreage relinquished	Same as Kuwait	Same as Kuwait	70% of unproductive acreage by 7th yr
Term of lease	Variable	Variable	Variable	Variable	Variable	Variable
Bonus	$1MM[c] within 60 days of lease; $2MM within 60 days of discovery.	Signature bonus—$25MM; 1 block out of 3 production bonus: $1, $2, and $3MM on 100M, 200M, and 300M B/D, respectively.	Signature £7MM; 4th anniversary of 100M B/D, £7MM; 500M B/D, £4MM.	None	Signature—$500,000; discovery—$1MM; production 70M B/D, $4MM.	None
Royalty	12½% of posted price	None	12½% of posted price with £2MM guarantee	20% of posted price	20% of posted price	20% of posted price, drops to 15% when 50% of taxable profits exceeds 20% of gross income.

Rental	None	$1.50 to $4.25/acre annually	£ 1MM/yr until discovery or £ 2MM/yr after 2nd yr until regular exports are achieved.	$2.5MM annually until discovery; $1.5MM/yr thereafter. In year of discovery up to $1.5MM may be credited against royalties.	$5/sq km annually for first 5 yr, rising to $500 in 26th yr.	Applied against royalties
Income tax	50% of taxable income	50% of taxable income	50% of taxable income	50% of taxable income	50% of taxable income	50% of taxable income
Exploration commitment	$13MM in 1st 8 yr	$12MM in 1st 4 yr	Must drill	2 rigs in continuous operation	$5MM within 2 yr	$10MM in 2 yr; $17.5MM next 10 yr.
Government participation	50%	50%	20%	10%	40%	75%

[a] After Kaufman,[3] courtesy of the SPE of the AIME.
[b] £ = Pound Sterling.
[c] MM = million.
(Note: agreements reached in 1971 with the producing countries may have altered the regulations presented here, i.e., the income tax was raised in many countries to 55%.)

TABLE 1-3

Summary of Offshore Regulations in Different Countries[a]

Regulation	Australia	Canada	Nigeria	Trinidad	United Kingdom[b]	Venezuela
Disposal system	Negotiated and competitive	Negotiated and competitive	Negotiated	Negotiated	Negotiated	Competitive service
Size of lease	Exploration- 400 to 10,000 sq mi; production- 125 to 150 sq mi.	Variable	10,000 sq mi exploration permit; 1000 sq mi prospecting license discretionary for production.	None	Production 250 sq km	Variable
Term of lease	Exploration- 6 yr, renewable in 5-yr increments. Production- 21 yr, renewable for an additional 21 yr.	Exploration, 3–4 yr, renewable 6 times for 1 yr each. Production- 21 yr from date of commercial production.	Exploration and prospecting- 4 yr. Production- 30 to 40 yr.	Exploration- 5 yr, renewable for 5 more. Production- 30 yr, renewable for 15 more.	Exploration- 3 yr; production- 6 yr, renewable for 40 yr.	5-yr development period; 15- to 20-yr production period from date of commercial production.

	Bonus	Rental
	Variable	Exploration-$5/block. Production $3,000/block.
	Variable	Production 1st yr 50¢/acre; 2nd yr and succeeding yr, $1/acre.
	$1.2 million/1000 sq mi during prospecting, plus same amount for production.	2 shillings/sq mi until discovery; 2 shillings 6 pence/acre to 10 shillings/acre thereafter.
	None	Annual exploration license 28¢/acre on leases under 1000 acres to 5¢/acre on leases over 100,000 acres. License is applied to exploration commitment. Mining license 56¢/acre 1st yr to $2.10 from 6th yr. Applied against royalties.
	£1,000/yr for exploration	Production license- 1st 6 yr, £25/sq km/yr. Rental 7th yr is £40/sq km, increasing £25/yr thereafter to max. of £290/sq km/yr. Applied against royalties.
	Government shares in profits relative to its investment. The govt. makes no investment until commercial production is established.	

TABLE 1-3 (continued)

Summary of Offshore Regulations in Different Countries[a]

Regulation	Australia	Canada	Nigeria	Trinidad	United Kingdom	Venezuela
Royalty	10% if 5 blocks selected, plus 1-2½% additional if 6 blocks selected.	5%	10% on oil and 4% on gas to 7 mi offshore; 8% on oil and 3.2% on gas farther out.	12% on oil and 1½% on gas	12½%	Government shares in profit relative to its investment. The government makes no investment until commercial production is established.
Income tax	Not available	Not available	50% of net income	50% of net income		
Exploration	Approved work program	Approved work program	Must drill 120,000 ft in 4 yr.	Negotiated	Negotiated	Negotiated

[a] After Kaufman,[3] courtesy of the SPE of AIME.
[b] £ = Pound Sterling.

(Note: agreements reached in 1971 with the producing countries may have altered the regulations presented here.)

Introduction to the Technology of Secondary Recovery

2. Sweep efficiency: the degree to which the reservoir will be contacted by the injected fluid.
3. Recovery efficiency: the thoroughness with which the process cleans the contacted rock of oil.

The oil saturation of the reservoir at any point in time is a function of the initial oil content, the effectiveness of the natural drive mechanism supporting the crude oil production, and the length of time the reservoir has been produced. The sweep efficiency is controlled by the location of the injectors and producers, reservoir heterogeneity, the mobility ratio, and the economic-limit oil production rate. The recovery efficiency is controlled by the capillary forces within the reservoir.

The more complex the reservoir, the greater the need to utilize the expertise of all segments of a company. The exploration personnel take the first steps in determining the characteristics of the reservoir. Too often, however, the results of their work either are not available to the production department responsible for the field development or, if available, are disregarded. Core, log, and fluid analyses represent only very small portions of the reservoir. A major problem is how to extrapolate these data into the reservoir. Here the use of the expertise of the exploration personnel is essential. The existence, magnitude, and distribution of the reservoir properties are controlled by the geologic processes that produced the reservoir rock.[4] A geologic study of the cores combined with a general knowledge of the geologic processes can lead to the proper extrapolation of the laboratory core data. Of particular importance is the recognition of the existence of directional permeability and the determination of the type of porosity system. The knowledge of the extent and types of non-uniformities present in the reservoir will help in the design of special tests to evaluate reservoir performance.[1] The lithologic, structural, and stratigraphic characteristics of the reservoir rock, combined with the production practices of the developers, will determine the type of drive mechanism which will control the primary recovery from the reservoir. Geologic characteristics which control the primary performance of a reservoir will also influence any attempt to improve the recovery.[5]

The solution gas-drive primary mechanism generally provides the most favorable conditions for the successful application of improved recovery techniques.[5] A reservoir may have a solution gas-drive mechanism because reservoir characteristics preclude the utilization of any other form of natural energy, or production rates are so high that the other natural energy sources are ineffective.[6]

The existence of a gas-cap drive or water drive, however, does not preclude some form of secondary recovery operation. Supplementary energy may be added to reduce the life of the reservoir or to correct inefficiencies in the primary drive mechanism.

Secondary solution, recrystallization, and fracturing make representative sampling (coring) and analyses particularly difficult in most carbonates.[7] Some carbonates have not experienced these secondary processes and have an intercrystalline and/or intergranular pore system which geometrically is similar to that of some sandstones. The flow behavior of these reservoirs may be similar to that of the sandstone reservoirs.[7] Other carbonates may have two or more pore systems and may exhibit a composite production performance which is a function of the physical properties of each system and their interrelationships.[8] The distribution and movement of fluids in carbonates, however, do follow the same basic principles which govern them in sandstone reservoirs. Only through combined laboratory and field work will the puzzle of the carbonate reservoir behavior be solved.[7] It must always be remembered that no study, regardless how rigorous, will yield a correct interpretation of the reservoir characteristics if the input data are wrong. In the analysis of a reservoir, it is better to use a simple model for which the data can be gathered with certainty than to use a complex model based on uncertain data.[9]

When the initiation of a secondary recovery project is considered, the dominant trapping mechanism which produced the reservoir must also be considered. Goolsby[10] has suggested the following classification system for trapping mechanisms: (1) structural, (2) stratigraphic, and (3) combination of structural and stratigraphic. The type of trap often dictates where injectors should be placed to achieve maximum recovery efficiency. Traps with moderate to high relief generally require peripheral or bottom-water injection and/or crestal gas injection. Traps with low to moderate relief are generally handled with pattern flooding. Additional comments on the impact of the reservoir trap type on the secondary recovery process design are presented in later chapters.

It has been pointed out that a good approach to the analysis of a carbonate reservoir involves both field and laboratory work. Preferably the analysis should be made early in the life of the reservoir, when the needed data can be obtained with reasonable accuracy. Economic and competitive considerations are often in conflict as to the types and quantity of data to be collected for use in future operations; therefore, a balance must be struck between current and future needs.[6] In the analyses of older fields, caution should be exercised in reviewing the data because the decisions on the allocation between current and future needs have already been made and cost-saving plans may have resulted in misleading or erroneous data. It is important to know under what operational conditions the data were collected, as the reservoir is responsive to these conditions. When data are being extrapolated, it is imperative that the operational characteristics of both the past and future periods are considered. If the operational characteristics change during the two periods, the extrapolation must be adjusted. Reservoir heterogeneity must be considered when dealing with production or test data.

Variations in crude properties should be noted. Changes in saturation pressure with depth of about 0.8 psi/ft have been noted in high-relief fields such as the Burgan Field in Kuwait and the Weber Sand of Rangely Field in Colorado.[11] Crude gravity variations of 0.7° API per 100 ft have been noted in the LL-370 area of the Bolivar Coastal Field, Venezuela, and the Fifth Zone at the Newhall-Potrero Field, California.[11] This type of variation will influence the accuracy of reserve calculations.

In the following chapters, the characteristics of the secondary recovery processes and their applications are reviewed. Particular attention is paid to the variations in process design and application caused by the differences in the types of pore systems of the carbonate reservoir rocks. The material is organized to promote the use of analogy, which can be a particularly powerful tool. Care must, however, be exercised to guarantee that only proper analogies are chosen.

The study of the performance of other fields helps to minimize the amount of time spent in developing the understanding of the new reservoir. Much can be learned from both successful and unsuccessful field applications of theories and processes. The rapid and early elimination of procedures and processes already shown inapplicable in similar type reservoirs saves time and money.

A study of many unsuccessful waterfloods indicates a similarity of problems, which would result in unsuccessful projects regardless of the type of secondary recovery process used. They may be summarized as follows[12, 13]:

Engineering Analysis

1. An erroneous estimate of the recovery factor resulting from unforeseen poor reservoir communication, reservoir fracture and/or fault systems, or an adverse permeability variation.
2. An erroneous estimate of the areal extent and net thickness of the reservoir.
3. Unexpected drainage of oil and gas from the lease owing to unforeseen drilling and completion of offset wells and/or reservoir conditions.
4. A high connate water saturation which results in an increased effective permeability to water and consequent water bypassing.
5. The poor mechanical condition of the surface producing equipment, which was not fully considered.

6. Improper plugging of old wells.
7. Underdesign of injection and/or producing facilities resulting in equipment being unable to handle the increased amounts of fluids.

Economic Difficulties

1. Insufficient capital to both develop the project and then wait for the often long-delayed payout.
2. Changes in proration and allowables.
3. Changes in the value of crude oil products.

The above-described difficulties serve to emphasize the importance of a thorough and searching analysis of every project prior to starting operations.

The study of process performance in other fields also provides data for the early recognition of fluid breakthrough, accelerated corrosion, and other operational problems that might occur. These problems and the degree to which they will appear are not theoretically predictable but must be considered fully prior to starting any secondary recovery project. Simple problems such as the mechanical condition of the surface equipment will directly affect the response of a proposed project. Other costs to consider would be those of handling increased production. Early recognition of these problems is required to avoid the loss of wells, equipment failure, or loss of the flood front control.

Application Ranges for Secondary Recovery Processes

Figures 1-1 and 1-2 are designed to serve as a rough screening device when considering the applicability of a secondary recovery process. Figure 1-1, presented originally by Roberts and Walker,[2] shows the range of crude oil gravity suitable for the application of various processes.

The success of the waterflooding of low-gravity crude oils in California suggests that the lower limit of the waterflood range shown in Fig. 1-1 may be dropped to 17° API. Surfactants, polymers, or CO_2 may be added to water to improve its ability to displace oil. Surfactants reduce the oil–water interfacial tension (see Appendix B on capillarity), polymers cut the mobility of water, and CO_2 reduces the viscosity of oil. The lower the API crude gravity, the greater the impact of these changes on recovery. For crude gravities above 35° API, the benefit of using these chemicals may be quite small. A range of 17° to 35° API appears to be appropriate for these improved waterflood methods.

Introduction to the Technology of Secondary Recovery 17

The depth and pressure ranges for the applicability of the secondary recovery processes are presented in Fig. 1-2. It should be noted that the depth range for a process, such as the slug-miscible displacement, can be extended by postponing the start of the project until the reservoir pressure has been decreased as a result of primary production. The depth limits for the conventional gas drive, dry gas-miscible drive, and waterflood are unspecified, as they may all be physically practical at 10,000 ft or deeper. Economics will dictate a variable cutoff point in different areas.

Reservoir structural characteristics also control the applicability of certain processes. Pattern waterflooding, dispersed gas injection (conventional or miscible), and the thermal processes require reservoirs of moderate dip (less than 15°) with fairly uniform pore structure. Crestal gas injection (conventional or miscible) or peripheral water injection are used in reservoirs having steep dips (greater than 15°) or reefs. The pore structure should again be reasonably uniform; however, it is not as critical as it is in the case of pattern or dispersed processes.

In situ combustion, steam injection, and CO_2 flooding have not been employed often enough in carbonate reservoirs to merit a detailed discussion, and, therefore, they are not presented in this book. In situ combustion and steam injection projects were reported for the Panhandle Field, Texas; however, no field data are presented.[14] A CO_2 flood is just getting started in the SACROC Unit of the Kelly–Snyder Field, Texas.[15] Additional information on these processes can be found in the following suggested references: 16 through 20 for CO_2 floods and 21 through 24 for thermal methods.

Questions and Problems

1-1. Discuss the factors that influence the decision as to which type of secondary recovery process should be selected.
1-2. List the three porosity type systems that are commonly found in carbonate reservoirs. How do these systems differ from one another?
1-3. Upon what factors does the effectiveness of a secondary recovery project depend?
1-4. Why would fluid and rock samples taken from one well in a carbonate reservoir not be truly representative of that reservoir? Briefly discuss the problems of gathering representative samples that might be anticipated for each of the three porosity systems.
1-5. Which types of secondary recovery processes could be considered for the following:

(a) Depth range of 500–2000 ft and oil gravity of 15° API.
(b) Depth range of 2000–3000 ft and oil gravity of 25° API.
(c) Depth range of 4000–7000 ft and oil gravity of 40° API.
(d) Depths greater than 7000 ft.

1-6. List the geologic and mechanical factors that have an influence upon the applicability of a secondary recovery process. Briefly state how the process may be affected.

1-7. What factors would a banker consider prior to granting a loan to a reputable small company to initiate a secondary recovery project?

References

1. Essley, P. L., Jr.: "What is Reservoir Engineering?" *J. Pet. Tech.* (Jan., 1965) 19-25.
2. Roberts, G. R., Jr. and Walker, S. W.: "Fluid Injection for Increased Oil Recovery", Section II—Paper 20 presented at the *5th World Pet. Cong.*, New York (1959).
3. Kaufman, A.: "International Offshore Leasing Practices," *J. Pet. Tech.* (March, 1970) 247-252.
4. Hewitt, C. H.: "How Geology Can Help Engineer your Reservoirs", *Oil and Gas J.* (Nov. 14, 1966) 171-178.
5. Wilson, W. W.: "Supplement to Influence of Geological Factors on Secondary Recovery of Oil", in: *Secondary Recovery of Oil in the United States*, 2nd Ed. API (1950) 211-213.
6. Torrey, P. D.: "Evaluation of Secondary Recovery Prospects", in: *Economics of Petroleum Exploration, Development and Property Evaluation*, Southwestern Legal Foundation, Prentice-Hall, Englewood Cliffs, New Jersey (1961) 190-213.
7. Craze, R. C.: "Performance of Limestone Reservoirs", *Trans.*, AIME (1950) **189**, 287-294.
8. Keller, W. O. and Morse, R. A.: "Some Examples of Fluid Flow Mechanism in Limestone Reservoirs", *Trans.*, AIME (1949) **186**, 224-234.
9. Rowan, G. and Warren, J. E.: "A Systems Approach to Reservoir Engineering: Optimum Development Planning", *J. Can. Pet. Tech.* (July–Sept., 1967) 84-94.
10. Goolsby, J. L.: "Here's the Relation of Geology to Fluid Injection in Permian Carbonate Reservoirs, West Texas", *Oil and Gas J.* (July 31, 1967) 188-190.
11. Grant, H. K.: "How Fluid-Property Variation in a High-Relief Oil Field Affects Material Balance Calculations", *Oil and Gas J.* (Aug. 24, 1959) 93-99.
12. Jackson, R. W.: "Why Some Waterfloods Fail", *World Oil* (March, 1968) 65-68.
13. Wright, F. F.: "A Report of Unsuccessful Waterfloods", *Proceedings of the Fourth Oil Recovery Conference*, Texas Petroleum Research Committee (1952) 48-58.
14. Neslage, F. J.: "Gas Injection in Dolomite Reservoir, West Pampa Repressuring Association Report", *Proceedings of the Third Oil Conference*, Texas Petroleum Research Committee (1951) 119-142.
15. (Editorial): "Soltex Gets Green Light on CO_2 Flood in West Texas", *Oil and Gas J.* (Jan. 19, 1970) 44-45.
16. Holm, L. W.: "Carbon Dioxide Solvent Flooding for Increased Oil Recovery", *Trans.*, AIME (1959) **216**, 225-231.

Introduction to the Technology of Secondary Recovery

17. Holm, L. W.: "CO_2 Requirements in CO_2 Slug and Carbonated Water Oil Recovery Processes", *Producer's Monthly* (Sept., 1963) 6-28.
18. Beeson, D. M. and Ortloff, G. D.: "Laboratory Investigation of the Water-Driven Carbon Dioxide Process for Oil Recovery", *J. Pet. Tech.* (April, 1959) 63-66.
19. de Nevers, N. H.: "A Calculation Method for Carbonated Water Flooding", *Soc. Pet. Eng. J.* (March, 1964) 9-20.
20. de Nevers, N. H.: "Carbonated Waterflooding—Is it a Lab Success and a Field Failure?" *World Oil* (Sept., 1966) 93-95.
21. Harvey, R. D.: "Nonlinear Thermal Expansion of Coarse-Grained Limestone", *Materials Research and Standards* (Nov., 1967) 502-506.
22. Marchant, L. C. and Menzie, D. E.: "Limestone Rock Changes by in-situ Combustion Temperatures", *Producer's Monthly* (Nov., 1966) 22-24.
23. Smith, R. C.: *Mechanics of Secondary Oil Recovery*, Reinhold, New York (1966).
24. (Editorial): "Thermal Recovery Processes", *Pet Trans.*, AIME *Reprint Series No.* 7.

CHAPTER 2

Planning a Secondary Recovery Project

Introduction

The first step to the realization of a successful secondary recovery project is adequate sampling of the reservoir rocks and fluids (coring, fluid sampling, production testing, etc.). The keeping of accurate production records and the correct interpretation of formation logs are of the utmost importance.

The second step is to use the available laboratory facilities to their greatest possible extent. The available reservoir rock and fluid samples should be studied exhaustively to develop as complete an understanding of their properties as possible. Published laboratory data are also important and useful as they can augment data developed for the specific reservoir. Appendixes C, D, and E present a summary of laboratory data on reservoir rock and fluid properties.

A pilot test was formerly considered the third step. Today, the third step is generally some form of computer simulation. This change has taken place due to the development of better computer hardware and more sophisticated software.

The pilot tests of the proposed processes constitute the fourth step. They provide the final checks on the compatibility of the reservoir characteristics with those of the processes. The fifth and last step is the economic analysis of those processes which are demonstrated to be applicable to the reservoir. The final decision as to which process to employ is based on the results of these analyses.

Collecting Data for Planning a Secondary Recovery Project

In 1950, the American Petroleum Institute published the 2nd Edition of *Secondary Recovery of Oil in the United States*.[1] A schedule of pertinent information for the study of secondary recovery possibilities, which was presented in that book, is also outlined here.

Planning a Secondary Recovery Project

It is impractical to expect to develop all the data contained in the following list. The data collected should, however, be as complete as possible. The list is supplied only as a starting point for the "systems study" of a reservoir or related group of reservoirs.

General Data

A. Name and title of investigator submitting report.
B. Date report was submitted.
C. Common name of field.
D. Location of field.
E. Property map of field showing lease ownership.
F. Source of all information.
G. Possible sources of additional information.
H. Area of field, in acres, for each producing horizon.
I. An individual field map for each producing horizon, showing location of all wells drilled. Abandoned wells owing to depletion or water encroachment, producing oil wells, producing gas wells, and dry holes should be indicated by conventional symbols.
J. Discovery date for each producing horizon.

Primary Development Practice in Field

A. Drilling: Cable tools or rotary, drilling time, drilling cost, size of hole drilled, and type of rig used; special difficulties encountered in drilling.
B. Average well spacing.
C. Complete casing record: Sizes used, average amount used, average amount pulled, average amount left in wells, and present condition. Note age and condition of reclaimed pipe.
D. Cementing practice.
E. Use of explosives in well completion: Quantity and type of explosive used, placement of shots in reference to top and bottom of reservoirs. Size of shells used and average cleanout time required. Type of tamping if tamped shots were used. Number of wells shot and general results of shooting.
F. Acid treatment in well completion: Quantity and type of acid used; method of placing; method of removal of spent acid; number of wells acidized, and general results of acidization.
G. Gun perforation of casing: Average number of shots and number of shots per foot; section of pay zone generally perforated; comparison of initial productions from gun-perforated wells and wells completed with open hole below casing or with perforated liner and screen.

H. Amount and size of tubing used.
I. Liners and screens used.
J. Gravel packing.

Petroleum Geology

A. All available well records, including electrical, radioactive, temperature, and other types of logs. Comment on completeness and reliability of records.
B. Stratigraphy: Typical well records and location of wells from which records have been obtained; well elevations; general information regarding various geologic formations encountered, such as their thickness, continuity, and lithology; convergence of formations in field, and any unconformities. Records of deep test wells are important.
C. Structure: Surface and subsurface structure maps; faults, fracture zones, crevices, etc.; relation of gas, oil, and water occurrence to structural conditions.
D. Name of formations encountered. Note whether they carry oil, gas, or water, and the amounts thereof, or whether they were dry.
E. Average depth of each formation: Average interval between formations and horizon markers recognized by drillers and paleontologists.
F. Average thickness of productive formations.
G. Variation in thickness of zones throughout field, and extent and location of any pronounced lenticularity.
H. Composition of limestone reservoirs. Note mode of occurrence of oil and gas: whether in interbedded sandstone lenses, solution cavities, coral reefs, oolitic zones, fractures or joints, or along bedding planes and formation contacts.
I. Nature of cap rock and of beds immediately underlying reservoir; hardness; presence of water; shale replacements in the top or bottom of sand; irregularities in contact of reservoir with overlying and underlying beds. Locate a suitable spot for casing seat or packer placement above the oil reservoir.
J. Lateral changes of reservoir throughout field. Note variations in thickness, porosity, permeability, saturation, and cementing material. Indicate if the limits of oil production are defined by thinning out of the pay zone, by variations in porosity, by gas-productive areas, or by water in the formation. Note presence of unconformities and disconformities.
K. Position of gas, oil, and water contacts in reservoir; pay zones and their thicknesses. Note initial gas production of oil wells and variations

Planning a Secondary Recovery Project 23

of same throughout field. Note any early production of water with either gas or oil. Determine the source of water and its composition.
L. Data on dry holes: Did producing formation pinch out or carry water?

Reservoir Characteristics and Behavior

A. Initial and present bottom-hole pressure. Pressure–volume relations; characteristics of production-decline curves.
B. Gas-liberation and oil-shrinkage data at various pressures. Analysis of produced gas and of gas after treating; deviation factors, either calculated or experimental.
C. Initial gas–oil ratios and accumulated gas–oil ratios. Gas–oil ratios at time of primary depletion.
D. Existence and extent of initial and present gas-cap zones; indications of gas-cap expansion.
E. Extent and effectiveness of natural water encroachment.
F. Substantiation of existence of gas-cap and water zone by electrical logs in case flushing of cores renders detection difficult.
G. Reservoir temperature.
H. Viscosity of oil under existing reservoir conditions of temperature, pressure, and amount of gas in solution.
I. Character of oil: Color, gravity, paraffin point, fluidity at low temperatures, and results of distillation test (oil samples for analysis should be identified as follows: sample No., well No., lease, company, field, date taken, and general remarks).
J. Chemical analysis of connate water. Water sample for analysis should be identified (sample No., well No., lease, company, field, and general remarks). The sample should be collected under conditions that preclude contact with air insofar as that is possible. Analysis for calcium, magnesium, potassium, barium, sulfur compounds, and the bicarbonate anion should be made as soon as possible. Extraction of pore water must be done at in situ temperature of rock. Observation in the field should be made whether the water is corrosive to casing, tubing, and other production equipment, and whether an odor of hydrogen sulfide is present.
K. Tendency of crude oil to emulsify or oxidize.
L. Presence of depleted gas pays and zones of high permeability which might prevent effective and uniform injection of gas or water into oil-bearing strata.
M. Data pertaining to possible pressure parting or other rock deformation under applied pressure.
N. Productivity indices.

Production Data

A. Total present daily oil production. Number of wells; present productive acres; individual lease and well daily oil and water production figures.
B. Potential tests of oil and water production, and method of obtaining same; production records of typical large, medium, and small wells, including date drilled, initial production, pressure, present oil and water production, gas–oil ratios, and date of abandonment; influence of gas production, variations in reservoir conditions, total production, and life of wells; reasons for any dry holes in field.
C. Gas-production history of field from strictly gas-bearing formations:
   1. Number of producing gas wells.
   2. Total present open-flow capacity.
   3. Average present rock pressure.
   4. Average gathering-line pressure.
   5. Average main-line pressure.
   6. Average age of gas wells.
   7. Original rock pressure.
   8. Original open-flow capacity.
   9. Total gas production by convenient time periods.
   10. Cumulative gas production.
   11. Estimated gas reserves, with particular attention to gas available for secondary recovery operations.
D. Water production of field, individual leases, and wells. Methods employed for disposal of water; determination of source of water either from leaking casing, improperly abandoned wells, or from some horizon producing oil.
E. Response of field to vacuum: Amount of vacuum applied and period of application; operation of gasoline recovery plants in connection with vacuum; effect of vacuum on gravity of oil; average gasoline recovery from casinghead gas; type of gasoline extraction used; disposition of stripped gas.
F. Record of all abandonments: Method of plugging; amount of casing pulled and amount of casing left in holes; junk left in holes; average primary production of wells at time of abandonment.
G. Estimated remaining primary reserves of field.
H. Results, if any, from secondary recovery operations. Estimated secondary reserves of field, based on results from existing projects or from core analyses data.

Primary Operating Practice in Field

A. Subsurface equipment in pumping wells: Anchor, working barrel, flood nipple, cups, balls and seats, rods, and production packers.

- B. Surface equipment for pumping wells: Pumping jacks, derricks or gin poles, receiving tanks, gas lines, oil lines, well-head connections, pull rods or pull lines, and central powers; oil storage facilities; condition of equipment.
- C. Pumping frequency: Time required to pump off; estimated lifting costs.
- D. Gas-lift installations: Type, pressure employed, and cost of operation.
- E. Turbine and hydraulic pumps: Comparison of results with conventional pumping equipment; cost of operation.
- F. Method of oil and water separation (treatment of emulsion).
- G. Average total cost of surface-well equipment.
- H. Common operating troubles:
  1. Paraffin and basic sediment.
  2. Scale deposition on sand, casing, and tubing.
  3. Casing and tubing leaks.
  4. Corrosion of equipment.
  5. Excessive wear and parting of rods.
  6. Abrasion of working barrel by floating sand.
  7. Accumulation of cavings and muck in bottom of hole.
- I. Cleanout procedure: Method employed, size of shot used, time required, average cost, and general results.
- J. Estimated increase in production by general rehabilitation of wells and equipment.
- K. Possibilities for rehabilitation of abandoned wells.
- L. Age of equipment.

Water Supply for Waterflooding Operations

- A. Surface supply: Name and location of streams; proximity to site of operations; approximate area of water shed; average annual rainfall in inches; stream flow during various seasons of the year; turbidity of water during various seasons of year; pollution, if any, of stream; corrosive nature of water; availability of dam sites; prior use of water for domestic and industrial supply, for carrying sewage, for hydroelectric power generation, or for irrigation.
- B. Subsurface supply: Depth, thickness, and productive capacity of subsurface water-bearing formations; turbidity of water; production, artesian flow or by pumping; effect of previous withdrawal of water on artesian pressure in flowing wells and on level of water table in pumping wells. Note all water-bearing horizons encountered in drilling oil wells.
- C. Water-sample data. Same as item J, p. 23.
- D. Water treatment required.

E. Mixing test of produced water and supply water.
F. Observation of precipitation rate or increase in turbidity of subsurface sample taken without exposure to air.
G. Estimated cost of water-supply development.

Gas Supply for Gas Repressuring

A. Availability of gas in field: By production from lease, by purchase from adjacent leases, or by purchase from gas company; amount of gas presently sold from field, name of company selling gas, and selling price.
B. Availability of gas outside of field:
    1. Ownership of gas.
    2. Purchaser of gas.
    3. Price paid per thousand cubic feet.
    4. Pressure base.
    5. Adequacy of supply for repressuring.
    6. Availability of any distress (flared, vented, etc.) gas.
    7. Proximity to pipelines.
    8. Length of pipeline required to make connection.
    9. Gathering-line pressure.
    10. Main-line pressure.
    11. Average closed-in rock pressure.
    12. Date of completion of wells.
    13. Maximum, minimum, and average open-flow capacity of wells.
    14. Age of wells.
    15. Original rock pressure at all sources.
    16. Recent production from wells.
    17. Estimated reserves available.
    18. Compressor facilities required.
C. Analysis of gas available for repressuring.
D. Possibility of using air for repressuring if gas supply is not adequate or is too costly.

Economic Considerations

A. Effect of topography on development cost, including effect on roads, pipelines, pumping equipment, well depth, etc.
B. Accessibility: Nearness to railroads, navigable streams, and highways.
C. Availability and cost of electric power.
D. State and pipeline proration practices (well allowables).
E. Laws and regulations governing improved recovery operations.
F. Name and address of operators with lease lists and acreage; description of leases. Probable productive acreage available for improved recovery

Planning a Secondary Recovery Project 27

operations, and cost of acreage; separation of deep rights from shallow rights; possible damage claims; existing lease terms; adverse lease terms; list of royalty owners and possibilities for unitization of royalty interests; willingness of operators to unitize and cooperate in recovery development; and ability of all operators to contribute proportionate share of cost.

G. Production, excise, and *ad valorem* taxes.
H. Availability of labor and wage scales. Nearness to supply houses.
I. Adaptability of recovery well-spacing pattern to lease boundary lines.
J. Markets for oil:
   1. Grade of oil.
   2. Pipeline facilities (condition and ownership).
   3. Condition of main storage tanks.
   4. Frequency with which oil is shipped.
   5. Ultimate purchaser of crude oil.
   6. Desirability of crude oil.
   7. Location and capacity of refinery connections.
   8. Outlook on future market for oil.
   9. Current price paid for crude and any premium.
   10. Transportation of oil by tank cars or tank trucks.
   11. Differences in transportation facilities during various seasons of year.
K. Capital cost of new equipment for improved recovery operations.
L. Variation in development cost based upon different volume and pressure requirements of repressuring media; variation in development cost based on type of process.
M. Increase in recovery owing to improved recovery techniques. Variation in recovery based on type of process; risk associated with each process.
N. Increase in operating expenses owing to improved recovery operations. Variation in expenses based on type of process.

Summary

Investigator must comment regarding the most applicable secondary recovery method, and desirability of further investigation and core testing of the producing formation.

Laboratory Investigation

Porosity, permeability, initial fluid saturations, capillary pressure, pore-size distribution, k_g/k_o and k_o/k_w relationships, saturation pressure, solution GOR, crude oil viscosity, and formation volume factor are among the data which should be developed for the reservoir by laboratory analysis. Of particular importance are the permeability and saturation relationships for the reservoir, which are treated in detail later in this chapter. Appendixes C, D, and E present summaries of published laboratory data on formation volume factors, viscosity, and compressibility, respectively. The list which follows was also included in the schedule of pertinent information for the study of secondary recovery possibilities.[1]

Core Analyses Data

A. Type of core: Cable-tool, side-wall, or chip; rotary with either diamond, wire-line, or conventional core-cutter heads; coring fluid used.
B. Names of organization or individuals who made analysis, and dates of coring, sampling, and analyzing.
C. Oil saturation data and profile:
  1. Discussion of any differences between saturation of specific cores and generally recognized oil saturation of reservoir.
  2. Oil content, in barrels per acre, and average oil content per acre-foot.
D. Water saturation data and profile:
  1. Determination of interstitial water content: Extent of infiltration of water from drilling fluid; indications, if any, of presence of water from accidental or intentional flood.
  2. Comparison of water content determined by core analyses and calculated from electric logs.
  3. Determination of water content by capillary displacement method.
E. Porosity data and profile.
F. Permeability data and profile.
G. Interpretation of core analyses data should include:
  1. Determination of continuity along bedding planes.
  2. Classification of formation in respect to saturation, porosity, permeability, and thickness.
  3. Determination of the minimum permeability that will be affected by secondary recovery operations at the economic limit of production.
H. Residual oil saturation after complete water drive, or after complete gas expansion, and rate of oil production as indicated by laboratory flooding and repressuring tests.
I. Relative permeability of reservoir to gas and to oil.

Testing of Secondary Recovery Processes

When one or more processes are being considered for application to a particular reservoir, laboratory analysis of the processes is essential. The tests should be run at as close to actual reservoir conditions as possible. Factors which should be considered are (1) the reactivity of the reservoir rock with waters of varying salinity, (2) compatibility of proposed injection fluids with reservoir fluids, (3) conditions required to achieve miscibility, (4) amount of crude oil burned during an in situ combustion trial, and (5) recoveries obtained by the various processes. Testing should also be directed toward proving the applicability of design procedures presented in the literature.

Effective Permeabilities

When two or three phases are present, the concept of permeability may be applied to each phase. The resulting parameter is the effective permeability, k_o, k_g, or k_w. It is known that an effective permeability is a function of both saturation and the distribution of the particular phase within the pore space. The sum of the effective permeabilities will always be less than the absolute permeability, k. This is caused by two factors: (1) when several phases occupy the pore space, each fluid has less cross-sectional area to flow through, and (2) the effective path length of flow for each fluid is increased.

Relative Permeabilities

Relative permeability is the ratio of effective permeability to absolute permeability. This may be written as follows:

$$k_{rg} = \frac{k_g}{k}; \qquad k_{ro} = \frac{k_o}{k}; \qquad k_{rw} = \frac{k_w}{k} \qquad (2\text{-}1)$$

It is sometimes advantageous to consider only the saturation range in which fluids are mobile. A nonmobile fluid, such as irreducible water, may be considered to be a part of the rock itself. When this is done, saturations are expressed as effective saturations and the corresponding relative permeability is obtained by normalizing effective permeabilities to that of the nonwetting phase value at the irreducible phase saturation. Relative permeability curves for sandstone and limestone cores studied by Craze[2] indicated no marked characteristic variations which could be attributed specifically to the formation composition. (See Appendix F for a general discussion of certain concepts of relative permeability.) The variations in the displacement of the curves were found to be a function of the textural differences and the pore size distribution for both. The shape of capillary pressure curves were found to be controlled in the same manner. Limestones generally have a wider variation in the shape of the pore spaces and distribution of pore sizes than sandstone reservoirs.

The relationship of permeability of a rock and its capillary desaturation curve has been recognized for many years. Purcell's[3] work led to the development of the following equation for the absolute permeability by Amyx et al.[4]:

$$k = 10.24(\sigma \cos \theta)^2 \phi a \int_{S=0}^{S=1} \frac{dS}{P_c^2} \qquad (2\text{-}2)$$

where:

k = absolute permeability, md.
ϕ = fractional porosity.
σ = interfacial tension, dynes/cm.
θ = contact angle.
P_c = capillary pressure, psi.
a = "lithology factor"—required to balance the equation.
S = fractional fluid saturation.

An expanded discussion of wettability and capillary pressure concepts is presented in Appendix B.

Wyllie and Gardner[5] derived a similar equation for absolute permeability, k, based on the Kozeny–Carman equation and Darcy's law for linear flow. Their equation can be presented as follows:

$$k = \frac{\phi(\sigma \cos \theta)^2}{2.5(\tau)_{S_w=1}} \int_0^1 \frac{dS_w}{P_c^2} \qquad (2\text{-}3)$$

In the above equation, τ is a tortuosity factor which accounts for the increased resistance to water flow caused by the interference offered by the hydrocarbon phase and the geometry of pore interconnections. An additional discussion of surface area and tortuosity concepts is presented in Appendix G.

The following equation for the relative permeability to water at a specific water saturation is derived from Eq. 2-3:

$$(k_{rw})_{S_w} = \frac{(\tau_w)_{S_w=1} \int_0^{S_w} (dS_w/P_c^2)}{(\tau_w)_{S_w} \int_0^1 (dS_w/P_c^2)} \qquad (2\text{-}4)$$

Equation 2-4 can be simplified by defining two relative factors: a relative tortuosity factor, τ_r, and a relative area factor, A_r. These are defined as follows:

$$(\tau_{rw})_{S_w} = \frac{(\tau_w)_{S_w=1}}{(\tau_w)_{S_w}} \tag{2-5}$$

$$(A_{rw})_{S_w} = \frac{\int_0^{S_w}(dS_w/P_c^2)}{\int_0^1(dS_w/P_c^2)} \tag{2-6}$$

Therefore,

$$(k_{rw})_{S_w} = (\tau_{rw})_{S_w} \cdot (A_{rw})_{S_w} \tag{2-7}$$

The relative permeability to the nonwetting phase (k_{rh}) can be similarly defined in terms of the tortuosity and relative area factors as follows:

$$(k_{rh})_{S_w} = (\tau_{rh})_{S_w} \cdot (A_{rh})_{S_w} \tag{2-8}$$

Burdine[6] expanded on Purcell's work and provided a method of graphical integration of the $1/P_c^2$ versus S_w curve.

The relative tortuosity factors were obtained by dividing measured relative permeabilities by appropriate calculated relative area factors. Burdine[6] found good correlation between τ_r and S_w. The relative tortuosity factors for water and hydrocarbons can be expressed in terms of wetting phase saturation as follows:

$$(\tau_{rw})_{S_w}^{1/2} = \frac{S_w - S_{wi}}{1 - S_{wi}} \tag{2-9}$$

$$(\tau_{rh})_{S_w}^{1/2} = \frac{1 - S_w - S_{hc}}{1 - S_{wi} - S_{hc}} = \frac{(S_M - S_w)}{(S_M - S_{wi})} \tag{2-10}$$

where S_{wi} = irreducible water saturation, S_{hc} = critical hydrocarbon saturation, and S_M = water saturation at which the hydrocarbon phase first attains hydraulic continuity $(S_M = 1 - S_{hc})$.

Equations for the relative permeabilities written in terms of water saturation and the capillary pressure are as follows:

$$(k_{rw})_{S_w} = \left(\frac{S_w - S_{wi}}{1 - S_{wi}}\right)^2 \frac{\int_{S_{wi}}^{S_w} (dS_w/P_c^2)}{\int_{S_{wi}}^{1} (dS_w/P_c^2)} \quad (2\text{-}11)$$

$$(k_{rh})_{S_w} = \left(\frac{S_M - S_w}{S_M - S_{wi}}\right)^2 \frac{\int_{S_w}^{1} (dS_w/P_c^2)}{\int_{S_{wi}}^{1} (dS_w/P_c^2)} \quad (2\text{-}12)$$

These equations can be used over the whole mobile range of water saturations (i.e., $S_{wi} < S_w < 1$). The relative permeability to water, k_{rw}, is normalized to k_w at $S_w = 1$. Both equations were developed by Burdine[6] to handle one hydrocarbon phase (either oil or gas) and interstitial water. These equations were simplified by Wyllie and Spangler[7] and Corey[8] to yield two-phase relative permeabilities as a function of saturation. Examination of relative tortuosity factor curves revealed that if $S_M = 1$, the saturation ($\approx 5\%$) of the hydrocarbon phase has not been reached, and[7]

$$(\tau_{rw})^{1/2} S_w + (\tau_{rh})^{1/2} S_w = 1 \quad (2\text{-}13)$$

The variance of predicted versus actual tortuosity as a result of this simplification becomes less important as the hydrocarbon phase saturation increases. Saturations may be expressed in terms of an effective saturation which for water, S_w^*, is defined as [8]

$$S_w^* = \frac{S_w - S_{wi}}{1 - S_{wi}} \quad (2\text{-}14)$$

Corey[8] has shown that in the case of some rocks, $1/P_c^2$ versus effective saturation may be considered a straight line over a considerable range of saturations. For such a linear relationship, the values of the integral ratios in Eqs. 2-11 and 2-12 become $(S_w^*)^2$ and $1 - (S_w^*)^2$, respectively.[8]

When these simplifications are made, the following equations are obtained for relative permeabilities:

$$(k_{rw})_{S_w} = (S_w^*)^2 (S_w^*)^2 = (S_w^*)^4 \quad (2\text{-}15)$$

$$(k_{rh})_{S_w} = (1 - S_w^*)^2 [1 - (S_w^*)^2] \quad (2\text{-}16)$$

Planning a Secondary Recovery Project

TABLE 2-1

Two-Phase Relative Permeabilities Expressed in Terms of S_w
(Water Saturations Determined by Drainage Rather than by Imbibition)

Type of rock	k_{rw}	k_{rh}
Unconsolidated, well-sorted oolites	$(S_w{}^*)^3$	$(1 - S_w{}^*)^3$
Cemented sandstone, oolitic limestone, and rocks with small vugs	$(S_w{}^*)^4$	$(1 - S_w{}^*)^2 [1 - (S_w{}^*)^2]$

Wyllie and Gardner[5] developed these important relative permeability–saturation relationships for various pore size distributions (capillary-pressure curves) which they believed to be characteristic of different types of rocks. These are given in Table 2-1 in terms of effective wetting phase saturations. The k_{rh} values calculated from relationships presented in Table 2-1 are normalized to effective hydrocarbon permeability at the irreducible water saturation. The k_{rw} values are normalized to absolute permeability.[5]

Three-phase relative permeability equations for preferentially water-wet systems operating where water and oil saturations are determined by the drainage cycle rather than by imbibition have also been given by Wyllie and Gardner[5] and are presented in Table 2-2.

The following factors should be taken in consideration when using the equations presented in Table 2-2.

1. The k_{rw} for water is normalized to absolute permeability.
2. The values of k_{ro} and k_{rg} calculated from these relationships are both normalized to the effective hydrocarbon permeability at irreducible water saturation. Inasmuch as they are normalized to the same base, k_{rg}/k_{ro} values may be calculated directly by using these equations. This is not true, of course, for water–hydrocarbon relative permeability ratios.
3. The gas and oil relative permeability equations do not include provision for residual oil saturation. When $S_w = S_{wi}$, k_{ro} is equal to $[S_o/(1 - S_{wi})]^4$ (for cemented sandstones, oolitic limestones, and vugular rocks). To handle residual oil saturation, this relationship should be altered to $[(S_o - S_{or})/(1 - S_{wi})]^4$. As a rule of thumb, S_{or} may be taken to be equal to 0.15.

TABLE 2-2

Three-Phase Relative Permeabilities Expressed in Terms of Saturations[a]

Type of rock	k_{rw}	k_{ro}	k_{rg}
Cemented sandstone, oolitic limestone, and vugular rocks	$(S_w{}^*)^4$	$\dfrac{S_o{}^3(2S_w + S_o - 2S_{wi})}{(1 - S_{wi})^4}$	$\dfrac{S_g{}^2[(1 - S_{wi})^2 - (S_L - S_{wi})^2]}{(1 - S_{wi})^4}$

[a] After Wyllie and Gardner.[5]

4. $S_w{}^* = \dfrac{S_w - S_{wi}}{1 - S_{wi}}$

5. $S_L = S_w + S_o$, where S_L = total liquid saturation, fraction.

Additional comments on relative permeability are presented in Appendix F.

Johnson[9] developed a simplified technique for solving Corey's equation for the relationship of the relative permeability ratio to gas saturation. Figure 2-1 illustrates the shapes for the Corey relative permeability ratio curves for several values of S_m and S_{LR}, where S_m = the lowest oil saturation fraction at which the gas tortuosity is infinite and S_{LR} = the residual liquid saturation, fraction. The S_m, which is a measure of the effect of stratification upon gas and oil relative permeabilities, was studied by Corey and Rathjens.[10] For uniform cores, it was found that S_m was essentially equal to one. When cores exhibited stratification perpendicular to the direction of flow, S_m exceeded one. On the other hand, when the stratification was parallel to the direction of flow, S_m was less than one.

Johnson[9] constructed Figs. 2-2 to 2-4 by assuming values of S_{LR}, S_m, and k_{rg}/k_{ro} and solving for $S_g(S_g = 1 - S_L)$ using Corey's equations. This calculation was repeated for various combinations of S_{LR} and S_m for k_{rg}/k_{ro} values of 10.0, 1.0, 0.1, 0.01, and 0.001. These figures can be used to develop a Corey curve that will fit and extrapolate experimental data. The span of the experimental data determines which one of the figures has to be used to determine S_{LR} and S_m which are constants for a particular rock–fluid system. If the data cover a wide range of permeability ratios, multiple determinations of S_{LR} and S_m can be made. If the calculated values differ from the experimental data, it means that there is no single Corey curve which would fit all the points exactly. The average of the values for each constant will yield a curve of reasonable fit. When an average is used, the variation in the constants should be kept in mind be-

Planning a Secondary Recovery Project

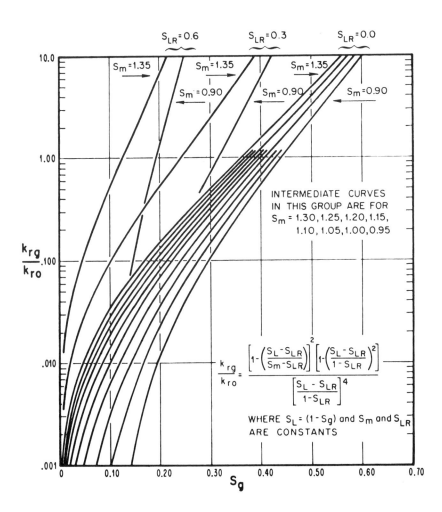

FIG. 2-1. Corey gas–oil relative permeability ratio curves for various values of S_g, S_{LR}, and S_m. (After Johnson,[9] courtesy of the SPE of AIME.)

cause the calculations in which these data are used may be very sensitive to the variations in different portions of the permeability ratio spectrum. Consequently, it may be desirable to use the constants derived for the restricted range used in the calculations instead of average values.

The steps which must be followed in using Figs. 2-1 to 2-4 are outlined below. Either one of these figures can be used first.

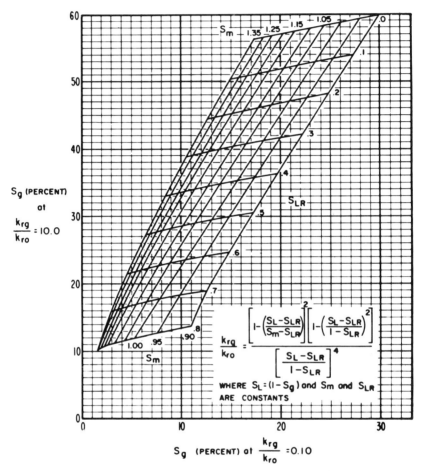

Fig. 2-2. Corey equation constants, S_{LR} and S_m, for various values of S_g at relative permeability ratios (k_{rg}/k_{ro}) of 10.0 and 0.10. (After Johnson,[9] courtesy of the SPE of AIME.)

1. Observe the range of the k_{rg}/k_{ro} values in the experimental data.
2. Determine the values of S_g for k_{rg}/k_{ro} ratios equal to 10.0, 1.0, 0.1, 0.01, and 0.001 if they lie within the range of the experimental data.
3. Choose a pair of k_{rg}/k_{ro} values, corresponding to the pairs used in constructing the figures. This choice determines which one of the figures should be used first.
4. Enter the first figure, using the S_g values corresponding to the pair of k_{rg}/k_{ro} values. The intersection of the two S_g values is also the inter-

Planning a Secondary Recovery Project

section of two unique, S_{LR} and S_m, constants. These two constants determine the character of the Corey curve.

5. Using the S_{LR} and S_m values from step 4, enter the other two figures and determine the rest of the S_g and k_{rg}/k_{ro} values for the Corey curve.

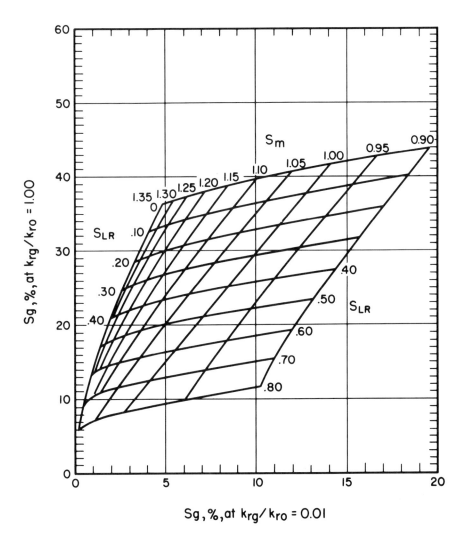

FIG. 2-3. Corey equation constants, S_{LR} and S_m, for various values of S_g at relative permeability ratios (k_{rg}/k_{ro}) of 1.0 and 0.01. (After Johnson,[9] courtesy of the SPE of AIME.)

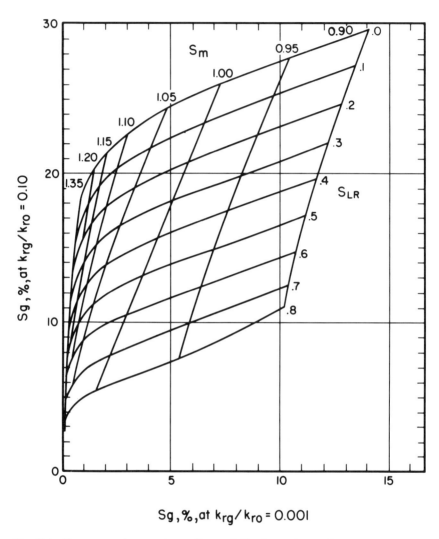

FIG. 2-4. Corey equation constants, S_{LR} and S_m, for various values of S_g at relative permeability ratios (k_{rg}/k_{ro}) of 0.10 and 0.001. (After Johnson,[9] courtesy of the SPE of AIME.)

Data Processing by Computer

The computer is becoming an increasingly important tool in the evaluation of the applicability of secondary recovery processes. Most of the predictive techniques discussed in the later chapters of this book have been "programmed" for computer processing. Use of the computer reduces the time, previously required to make the calculations by hand or calculator, from days or hours to minutes or seconds. This reduction in time provides the additional benefit to engineers of being able to study the effects of any or all the major variables which previously were not considered because of the long time required. Material balance analysis, Tarner[11] prediction of solution gas-drive recovery, and waterflood recovery predictions using the Buckley–Leverett[12] or Dykstra–Parsons[13] techniques are but a few of the more routine calculations for which programs have been developed.

The computer must always be supplied with reliable data. There is a basic truth which underlies all computer work: "garbage in = garbage out." When computer analysis is used as the third step in flood design, the first and second steps become doubly important and the "key" word to be kept in mind all through steps 1 and 2 is "representative." Inadequate sampling in the case of carbonate reservoirs can be a major problem.

Within the past few years, "reservoir simulation" has become the practical tool of the reservoir engineer. Although it is an old tool, advances in computer hardware have resulted in flexibility not previously permitted in programming. Odeh[14] points out that simulation is the representation of a process by either a theoretical or physical model. A secondary recovery reservoir simulator may utilize a volumetric, material balance, or frontal advance model. It is important to realize that any model constructed will yield answers that are limited by the type of basic model selected to describe fluid movement.

Simulation is a valuable tool in studying secondary recovery processes because it permits the observer to study a variety of field performance schemes over a wide range of operating conditions. Coats[15] presented the following list of problems which could be solved by simulation:

1. Determination of the performance of an oil field under water injection, gas injection, or natural depletion.
2. Comparison of the advisability of flank waterflooding as opposed to pattern waterflooding.
3. Determination of the effects of well location and spacing.
4. Estimation of the effect of producing rate on recovery.
5. Calculation of the total gas field deliverability for a given number of wells at certain specified locations.
6. Estimation of the lease-line drainage in heterogeneous oil or gas fields.

This list presents only a few of the possible uses of simulation. Coats[15] presented a good general rule to follow in the selection of a simulator:

"Select the least complicated model and grossest reservoir description that will allow the desired estimation of reservoir performance." For additional information on reservoir simulation, the reader is referred to the articles by Odeh[14] and Coats.[15]

Pilot Tests and Their Value

The purpose of a pilot test is to determine the response of a reservoir to secondary recovery process prior to the investment of large sums of capital required to develop the full-scale project. Inasmuch as a successful pilot test usually leads to expansion to a full-scale operation, the design and location of a pilot is critical. Three important rules should be followed in planning a pilot test[16]:

1. The location of the pilot test should be within the area of the final full-scale flood.
2. The area selected for the pilot test should be representative of the entire reservoir. If there is no such one area, then several areas should be selected. The areas selected should have wells that are in good mechanical condition.
3. The expected performance on the basis of which the pilot test is to be expanded should be specified before starting the test. The objectives and desired performance to meet them should be recorded as part of the pilot plan.

The first two rules are particularly important in carbonate reservoirs with highly variable porosity systems.

The results of pilot tests must be analyzed with care. A pilot test site generally lies within an "unconfined" area, whereas the total reservoir lies in a "confined" area. In an "unconfined" area, oil can be gained or lost to an adjacent portion of the reservoir. Also, only a portion of the fluid injected is useful in moving oil to the producing wells of the pilot test area; the rest of the fluid escapes to the surrounding reservoir. These important factors can be compensated for by proper analysis and interpretation of the pilot test results. Fisher-Rosenbaum *et al.*[17] have noted that when the response of a pilot test is earlier than that predicted for a homogeneous reservoir, the behavior observed is a measure of the reservoir's heterogeneity. If the response occurs later than predicted, the formation permeability near the producing wells is possibly lower than the average formation permeability. It can also be the result of plugging or the presence of thief sands which carry the water to other portions of the reservoir.

Planning a Secondary Recovery Project

A pilot test determines the applicability of a proposed process or processes to a particular reservoir. "Piloting" of two processes in a single large reservoir is becoming more common. A pilot test may serve to determine the following processes:

1. The reservoir injectivity, which must be known when designing injection facilities.
2. The existence of directional permeability. In selecting injection points for the full-scale flood, one should take into consideration any preferential movement of the injected fluid.
3. The existence of thief zones which could cause the wholesale bypassing of crude oil and the failure of a proposed flood. If a thief zone is found, the feasibility of shutting it can be investigated.
4. The existence of any barriers to flow such as faults.
5. The existence of production problems caused or aggravated by the injection, e.g., corrosion, production of fines or sand, and emulsion formation.
6. The feasibility of converting producers to injectors.

It should be emphasized that the primary purpose of a pilot test should not be to determine the recovery factor or percentage of oil recovery from the reservoir. On studying pilots in several carbonate reservoirs, Buckwalter[18] found that the recovery indicated by pilot floods varies from a few percent to more than 150% of that of primary oil production. In most cases, however, his studies of waterfloods showed that full-scale floods did not have the same recovery factors as those indicated by the pilot tests due to wide variations in geologic factors; the pilot test areas were not representative of the whole reservoir. At best, pilot floods indicate floodability and yield a "ball-park" recovery factor.

Injection Profile Control

Detection of a thief zone in a reservoir being considered for secondary recovery is of prime importance. The economic success of any secondary recovery process is dependent upon the vertical and horizontal efficiency of the displacement process. If thief zones are present, they could lead to the cycling of excessive amounts of the injection fluid and might result in a failure of the project owing to poor sweep efficiency.

In the absence of major thief zones, it is still necessary to evaluate the overall injection profile. Any portion of the reservoir which allows entry of a large percentage of the injected fluid could later develop into a thief zone. Control of the injection profile has been achieved with varying degrees of success by selective perforation, selective plugging, or by reducing the mobility of the injected fluid relative to the crude oil.[19]

Selective plugging in injection wells is not new to the petroleum industry. Cement, colloidal clays, inert solids, paraffins, waxes, organic resins, emulsions, and silicic acid gels have all been utilized. Two common techniques of selective plugging are (1) the deposition of an agent that plugs the formation around the wellbore of the injector, and (2) the deposition of an agent that penetrates the formation and plugs the thief zone at a distance from the wellbore. The successful use of these agents depends upon the continuity of the impermeable layers throughout the reservoir. Without this, the injected fluid will reenter the thief zone once it has past the plugged zone. The vertical migration of fluid has likely been the cause of the failure of many attempts to control an injection profile.[19]

Where impermeable layers are not continuous, oil-in-water emulsions (having a high oil concentration), gels, and polymers have been utilized. The advantage of using these products is that they easily penetrate the formation and lodge up to 40 ft away from the injector. The volume of these materials used should be large enough to resist movement which may be caused by the differential pressure across the slug.

In a waterflood, the mobility ratio may be reduced by inclusion of oil-in-water emulsions or polymers in the injected water. These additives are similar in that they divert the flood from the channels having higher permeability toward the lower permeability zones.

Questions and Problems

2-1. List and discuss the principal types of data needed on an undeveloped field for future use in a secondary recovery project.

2-2. As mentioned in the chapter, laboratory data are important in helping one ascertain the probable success of a proposed project. What steps should be taken to ensure that the data are representative, without spending excessive sums of money?

2-3. Calculate the two-phase and three-phase relative permeability relationships for an oolitic limestone if the irreducible water saturation is 21% and cores indicate a water saturation of 35%.

2-4. Utilizing Johnson's technique, prepare a k_{rg}/k_{ro} versus S_g plot for the following laboratory data:

k_{rg}/k_{ro}	S_g
1.9	0.50
0.109	0.30

Planning a Secondary Recovery Project

Briefly discuss the factors which should be considered prior to directly using the computed curve in the field.

2-5. What types of reservoir problems may be solved by computer simulation?

2-6. What information is (a) given by a pilot test and (b) not given by a pilot test? Discuss the type of information.

2-7. If vertical movement of fluids occurs within the formation, would a plugging agent that plugs only the formation around the wellbore be successful? Why?

2-8. In waterflooding, should surfactants make contact angle less or greater than 90°? Explain why! (Additional information can be found in Appendix B.)

2-9. Describe the role of capillary forces in the displacement of oil by water.

2-10. With increasing content of carbonate particles in calcareous sandstones, would relative permeability curves be higher or lower than those for pure sandstones (if all other variables are considered the same). Explain for both k_{ro} and k_{rw}! (Additional information can be found in Appendix F.)

2-11. In waterflooding operations of carbonate rocks having fracture–matrix type porosity systems, should surfactants make the walls of fractures water- or oil-wet? (Additional information can be found in Appendix B.)

References

1. Torrey, P. D.: "Schedule of Pertinent Information for Study of Secondary-Recovery Possibilities", in: *Secondary Recovery of Oil in the United States*, 2nd Ed., API (1950) 109–114.
2. Craze, R. C.: "Performance of Limestone Reservoirs", *Trans.*, AIME (1950) **189**, 287–294.
3. Purcell, W. R.: "Capillary Pressures—Their Measurement Using Mercury and the Calculation of Permeability Therefrom", *Trans.*, AIME (1949) **148**, 39–48.
4. Amyx, J. W., Bass, D. M., Jr., and Whitting, R. L.: *Petroleum Reservoir Engineering—Physical Properties*, McGraw-Hill, New York (1960).
5. Wyllie, M. R. J. and Gardner, G. H. F.: "The Generalized Kozeny–Carman Equation, Its Application to Problems of Multiphase Flow in Porous Media", *World Oil* (March, 1958) 121–126 and (Apr., 1958) 220–228.
6. Burdine, N. T.: "Relative Permeability Calculations from Pore Size Distribution Data", *Trans.*, AIME (1953) **198**, 71–78.
7. Wyllie, M. R. J. and Spangler, M. R.: "Application of Electrical Resistivity Measurements to Problem of Fluid Flow in Porous Media", *Bull.*, AAPG (1952) **36**, 359–403.

8. Corey, A. T.: "The Interrelation between Gas and Oil Relative Permeabilities", *Producer's Monthly* (Nov., 1954) 38–41.
9. Johnson, C. E., Jr.: "Graphical Determination of the Constants in the Corey Equation for Gas-Oil Relative Permeability Ratio", *J. Pet. Tech.* (Oct., 1968) 1111–1113.
10. Corey, A. T. and Rathjens, C. H.: "Effect of Stratification on Relative Permeability", *Trans.*, AIME (1956) **207**, 358–360.
11. Tarner, J.: "How Different Size Gas Caps and Pressure Maintenance Programs Affect the Amount of Recoverable Oil", *Oil Weekly* (June 12, 1944) **114**, 32.
12. Buckley, S. E. and Leverett, M. C.: "Mechanism of Fluid Displacement in Sands", *Trans.*, AIME (1942) **146**, 107–116.
13. Dykstra, H. and Parsons, R. L.: "The Prediction of Oil Recovery by Waterflood", in: *Secondary Recovery of Oil in the United States*, 2nd. Ed., API (1950) 160–174.
14. Odeh, A. S.: "Reservoir Simulation ... What is it?" *J. Pet. Tech.* (Nov., 1969) 1383–1388.
15. Coats, K. H.: "Use and Misuse of Reservoir Simulation Models", *J. Pet. Tech.* (Nov., 1969) 1391–1398.
16. Smith, R. C.: *Mechanics of Secondary Recovery*, Reinhold, New York (1966).
17. Fisher-Rosenbaum, M. J. and Matthews, C. S.: "Studies on Pilot Water Flooding", *Trans.*, AIME (1959) **216**, 316–323.
18. Buckwalter, J. H.: "Some Pilot Waterflood Results in Carbonate Reservoirs", A Panel Discussion Presented in *Proceedings of the Southwest Petroleum Short Course* (1966) 205.
19. Robertson, J. O., Jr. and Oefelein, F. H.: "Plugging Thief Zones in Water Injection Wells", *J. Pet. Tech.* (Aug., 1967) 999–1004.

CHAPTER 3

General Principles of Waterflood Design

Introduction

An excellent comparison of the various waterflood prediction techniques[1-43] was presented by Schoeppel[44] in the *Oil and Gas Journal*. Table 3-1 summarizes the various prediction techniques presented, whereas Tables 3-2 to 3-4 provide details for the primary techniques summarized in Table 3-1. Major assumptions, data requirements, and comparisons of methods are presented in these tables. Tables 3-5 to 3-7 provide detailed comparisons of alternate predictive techniques within the major groupings. Reviewing these tables early in the life of a field provides direction to the future gathering of data.

Waterflooding "Rules of Thumb"

When considering the use of waterflooding for a specific reservoir, it is often valuable to get a quick look at the overall project economics. With this in mind, the following "rules of thumb" were accumulated by the authors. Two primary sources were used: N. van Wingen's lectures at the University of Southern California and "Practical Waterflooding Shortcuts" published in *World Oil*.[45]

1. Water requirements: $1\frac{1}{2}$ to 2 pore volumes.
2. Typical injection rates for wells average from 5 to 10 B/D/ft of reservoir for pattern flooding; a rate of 3 B/D/ft is considered minimum. A higher rate of 10 to 20 B/D/ft can be anticipated for aquifer injection.
3. Injection rates for sandstone reservoirs vary from 0.35 to 1.5 bbl/acre-ft with an average of 0.5 bbl/acre-ft. Limestone rates should run two to three times the average.
4. Response to flood can be expected when two thirds of the fill-up has been achieved (i.e., when two thirds of the voidage created by primary production is filled by the water injected).

TABLE 3-1

Classification of 33 Waterflood Prediction Methods[a]

Basic method	Modification

I. Methods primarily concerned with permeability heterogeneity-injectivity
   1. Dykstra–Parsons (1950)[1]
          (a) Johnson (1956)[2]
          (b) Felsenthal–Cobb–Heuer (1962)[b 3,4]
   2. Stiles (1949)[5]
          (a) Schmalz–Rahme (1950)[c 6]
          (b) Arps ("Modified Stiles") (1956)[7]
          (c) Ache (1957)[8]
          (d) Slider (1961)[9]
   3. Yuster–Suder–Calhoun (1949)[10]
          (a) Muskat (1950)[11]
          (b) Prats et al. (1959)[d 12]
   4. Prats–Matthews–Jewett–Baker (1959)[12]

II. Methods primarily concerned with areal sweep efficiency
   1. Muskat (1946)[13]
   2. Hurst (1953)[14]
   3. Atlantic–Richfield (1952–1959)[15,16,17,18]
   4. Aronofsky (1952–1956)[19,20]
   5. Deppe–Hauber (1961–1964)[21,22]

III. Methods primarily concerned with the displacement process
   1. Buckley–Leverett (1942)[23]
          (a) Terwilliger et al. (1951)[24]
          (b) Felsenthal–Yuster (1951)[25]
          (c) Welge (1952)[26]
          (d) Craig–Geffen–Morse (1955)[d 27,28]
          (e) Roberts (1959)[29]
          (f) Higgins–Leighton (1960-1964)[d 30,31,32,33]
   2. Craig–Geffen–Morse (1955)[27,28]
          (a) Hendrickson (1961)[34]
   3. Higgins–Leighton (1960–1964)[30,31,32,33]

IV. Miscellaneous theoretical methods
   1. Douglas–Blair–Wagner (1958)[35]
   2. Hiatt (1958)[36]
   3. Douglas–Peaceman–Rachford (1959)[37]
   4. Naar–Henderson (1961)[38]
   5. Warren–Cosgrove (1964)[39]
   6. Morel–Seytoux (1965)[40]

V. Empirical methods
   1. Guthrie–Greenberger (1955)[41]
   2. Schauer (1957)[42]
   3. Guerrero–Earlougher (1961)[43]

[a] After Schoeppel,[44] courtesy of *Oil and Gas Journal*.
[b] Also applies to Stiles method.
[c] Also applies to Yuster–Suder–Calhoun and Schauer methods.
[d] Also concerned with areal sweep problem. Also recognized as basic method.

TABLE 3-2

Characteristics of the Perfect Waterflood Prediction Method[a]

Effect	Characteristic
Fluid flow	Initial gas saturation is considered
	Saturation gradient is considered
	Varying injectivity is considered
Pattern	Applies to linear systems
	Applies to five-spot pattern
	Applies to other patterns
	Applicable to all mobility ratios
	Considers areal sweep
	Considers increased sweep after breakthrough
	Does not require published laboratory data
	Does not require additional laboratory data
Heterogeneity	Considers stratified reservoirs
	Considers crossflow
	Considers spatial variations

[a] After Schoeppel,[44] courtesy of *Oil and Gas Journal*.

5. Peak oil production rate is reached at the time of fill-up.
6. Gross production rate equals 80% of water injection rate. The rest of the injected water is lost outside of the patterns or to the aquifer.
7. Approximately one half of the secondary recovery oil (primary recovery oil left at time of flood plus increase in recovery) will be recovered prior to reaching peak production.
8. Incremental recovery due to waterflooding is equal to the primary for crude oils having gravities above 30° API. For lower crude gravities (15°–30° API), the incremental recovery ranges from 50% to 100% of primary recovery.
9. The analysis of many pattern floods indicates[46]:
   (a) areal efficiency = 70 to 100%,
   (b) vertical efficiency = 40 to 80%,
   (c) combined efficiency = 28 to 80% (median of 60%), and
   (d) residual oil saturation = 15 to 30%.
10. Anticipated production expenses of secondary recovery oil in the United States are around $1.00/bbl. Capital investment for plant and facilities is around $0.20/bbl.

TABLE 3-3

Basic Assumptions in Waterflood Prediction Methods
(After Schoeppel,[44] courtesy of Oil and Gas Journal.)

			(1)	(2)	(3)
				Fluid-Flow Effects	
				Considers	
	Method and Modification*	Date Presented	Initial gas saturation	Saturation gradient	Varying injectivity
	THE PERFECT METHOD		Yes	Yes	Yes
I.	Dykstra-Parsons[1]	1950	Yes	No	No
	Johnson[2]	1956	Yes	No	No
	Felsenthal-Cobb-Heuer[3,4]	1962	Yes	No	No
	Stiles[5]	1949	No	No	No
	Schmalz-Rahme[6]	1950	No	No	No
	Arps[7]	1950	No	No	No
	Ache[8]	1957	No	No	No
	Slider[9]	1961	Yes	No	Yes
	Yuster-Suder-Calhoun[10]	1949	Yes	No	Yes
	Muskat[11]	1950	No	No	Yes
	Prats et al.[12]	1959	Yes	No	Yes
II.	Muskat[13]	1946	No	No	No
	Hurst[14]	1953	No	No	No
	Atlantic-Richfield[15,16,17,18]	1952-59	No	No	Yes
	Aronofsky[19,20]	1952-56	No	No	Yes
	Deppe-Hauber[21,22]	1961-64	Yes	No	Yes
III.	Buckley-Leverett[23]	1942	No	Yes	No
	Welge[26]	1952	No	Yes	No
	Roberts[29]	1959	No	Yes	No
	Craig-Geffen-Morse[27,28]	1955	Yes	Yes	Yes
	Higgins-Leighton[30,31,32,33]	1960-64	Yes	Yes	Yes
	Hendrickson[34]	1961	No	Yes	No
IV.	Douglas-Blair-Wagner[35]	1958	No	Yes	Yes
	Hiatt[36]	1958	No	Yes	No
	Douglas et al.[37]	1959	No	Yes	Yes
	Naar-Henderson[38]	1961	No	Yes	No
	Warren-Cosgrove[39]	1964	No	Yes	No
	Morel-Seytoux[40]	1965	No	Yes	Yes
V.	Schauer[42]	1957	Yes	No	No
	Guerrero-Earlougher[43]	1961	Yes	No	No

* See footnote at the end of Table.

(4)	(5)	(6)	(7)	(8)	(9)	(10)	(11)
				Pattern Effects			

	Applies to			Considers		Requires	
Linear system	5-spot pattern	Other patterns	Mobility ratio used	Areal sweep	Increased sweep after break-through	Pub-lished lab data	Addi-tional lab data
Yes	Yes	Yes	Any	Yes	Yes	No	No
Yes	No	No	Any	No	No	No	No
Yes	No	No	Any	No	No	No	No
Yes	No	No	Any	No	No	No	No
Yes	No	No	1.0	No	No	No	No
Yes	No	No	1.0	No	No	Yes	No
Yes	No	No	1.0	No	No	No	No
—	Yes	No	1.0	No	No	No	No
Yes	Yes (?)	No	1.0	No	No	No	No
—	Yes	No	1.0	No	No	No	No
Yes	No	No	Any	No	No	No	No
—	Yes	No	Any	Yes	Yes	Yes	Yes
—	Yes	Yes	1.0	Yes	No	No	No
—	Yes	No	1.0	Yes	No	No	No
—	Yes	Yes	Any	Yes	Yes	Yes	No
—	Yes	No	Any	Yes	No	Yes	No
—	Yes	Yes	Any	Yes	Yes	Yes	No
Yes	No	No	—	No	No	No	No
Yes	No	No	—	No	No	No	No
Yes	No	No	—	No	No	Yes	No
—	Yes	No	Any	Yes	Yes	Yes	No
—	Yes	Yes	Any	Yes	Yes	Yes	No
—	Yes	No	Any	Yes	Yes	Yes	No
—	Yes	Yes	Any	Yes	Yes	Yes	No
Yes	No	No	—	No	No	No	No
—	Yes	Yes	Any	Yes	Yes	No	No
—	Yes	No	Any	Yes	Yes	Yes	No
Yes	No	No	—	No	No	No	No
—	Yes	Yes	Any	Yes	Yes	No	No
—	Yes	No	Any	Yes	No	No	No
—	Yes	Yes	—	No	No	No	No

TABLE 3-3 (continued)

(After Schoeppel,[44] courtesy of Oil and Gas Journal.)

	Method and Modification*	Date Presented	(12) Stratified reservoirs	(13) Crossflow	(14) Spatial variations
			Heterogeneity Effects		
			Considers		
	THE PERFECT METHOD		Yes	Yes	Yes
I.	Dykstra-Parsons[1]	1950	Yes	No	No
	Johnson[2]	1956	Yes	No	No
	Felsenthal-Cobb-Heuer[3,4]	1962	Yes	No	No
	Stiles[5]	1949	Yes	No	No
	Schmalz-Rahme[6]	1950	Yes	No	No
	Arps[7]	1956	Yes	No	No
	Ache[8]	1957	Yes	No	No
	Slider[9]	1961	Yes	No	No
	Yuster-Suder-Calhoun[10]	1949	Yes	No	No
	Muskat[11]	1950	Yes	No	No
	Prats et al.[12]	1959	Yes	No	No
II.	Muskat[13]	1946	No	No	No
	Hurst[14]	1953	Yes	No	No
	Atlantic-Richfield[15,16,17,18]	1952–59	Yes	No	No
	Aronofsky[19,20]	1952–56	Yes	No	No
	Deppe-Hauber[21,22]	1961–64	Yes	No	No
III.	Buckley-Leverett[23]	1942	No	No	No
	Welge[26]	1952	No	No	No
	Roberts[29]	1959	Yes	No	No
	Craig-Geffen-Morse[27,28]	1955	Yes	No	No
	Higgins-Leighton[30,31,32,33]	1960–64	Yes	No	No
	Hendrickson[34]	1961	Yes	No	No
IV.	Douglas-Blair-Wagner[35]	1958	Yes	No	No
	Hiatt[36]	1958	Yes	Yes	No
	Douglas et al.[37]	1959	Yes	No	No
	Naar-Henderson[38]	1961	Yes	No	No
	Warren-Cosgrove[39]	1964	Yes	Yes	No
	Morel-Seytoux[40]	1965	Yes	No	No
V.	Schauer[42]	1957	Yes	No	No
	Guerrero-Earlougher[43]	1961	Yes	No	No

* Categorized according to classification presented in Table 3-1.

Flood Design by Analogy

The study of existing and completed waterflood projects and natural water-drive reservoirs provides the basis for the extension of the "rules of thumb" approach to flood design. A study of the reservoirs in the Denver Basin (Nebraska and Colorado) showed that natural water-drive reservoirs in the Nebraska portion of the basin had primary recoveries of 40 to 45% of the initial stock tank oil-in-place.[47] In the Colorado portion of the basin, 18% was considered a good primary recovery (solution gas-drive mechanism) in similar types of reservoirs. The operators concluded that waterflooding would be feasible in the Colorado portion of the basin and that the ultimate recovery for a field could reasonably be expected to be 40%. The predicted performance of the West Lisbon waterflood in Louisiana was based on the past performance of the waterflood of the Southwest Lisbon Pettit reservoir, Louisiana. It was estimated that the recovery would be raised from 14% under primary depletion to 32% under flood.[48]

Callaway[49] suggested that by relating the results obtained from a waterflood to the reservoir parameters, which control the performance, a reliable set of experience factors can be obtained and the uncertainties can be greatly reduced. He divided the engineering factors involved in evaluating waterflood recovery into two sets of variables: (1) "primary variables," which are those factors bearing a direct mathematical relation to the amount of oil to be recovered; (2) "secondary variables," which operate indirectly through the primary variables to influence the oil recovery. The primary variables are (1) primary recovery efficiency, (2) connate water saturation, (3) sweep efficiency, (4) residual oil saturation, and (5) crude shrinkage. The secondary variables and the corresponding primary factors influenced (numbers in parentheses) are (a) oil viscosity (1, 3, 4); (b) permeability (1, 3, 4); (c) structural considerations (1, 3); (d) uniformity of reservoir rock (3); (e) type of flood (3); (f) time for start of flood (5); and (g) economic factors (1, 3, 4).

Callaway's[49] equation for evaluating the recovery by waterflooding is

$$WR = 7758\phi \left(\frac{1 - S_w}{B_{oi}}\right)\left[1 - E_p - \frac{B_{oi}}{B_o}\left[1 - E_o\left(1 - \frac{S_{or}}{1 - S_w}\right)\right]\right] \quad (3\text{-}1)$$

where:

WR = waterflood recovery, bbl/acre-ft.
B_{oi} = original formation volume factor for oil, bbl/STB.
B_o = formation volume factor for oil during waterflood operations, bbl/STB.

TABLE 3-4

Data Required for Waterflood Prediction Methods
(After Schoeppel,[44] courtesy of Oil and Gas Journal.)

Method and Modification*	Absolute permeability	Stratified bed thickness	Effective oil and water permeability	Relative permeability saturation curve
1. Dykstra-Parsons[1]	X	X	X	—
Johnson[2]	X	X	X	—
Felsenthal-Cobb-Heuer[3,4]	X	X	X	—
2. Stiles[5]	X	X	X	—
Schmalz-Rahme[6]	X	X	X	—
Arps[7]	X	X	X	X
Ache[8]	X	X	X	X
Slider[9]	X	X	X	X
3. Yuster-Suder-Calhoun[10]	X	X	**	—
Muskat[11]	X	X	X	—
Prats et al.[12]	X	X	X	X
4. Muskat[13]	—	—	—	—
Hurst[14]	X	X	—	—
Atlantic-Richfield[15,16,17,18]	X	X	X	—
Aronofsky[19,20]	—	—	—	—
5. Buckley-Leverett[23]	—	—	—	X
Welge[26]	—	—	—	X
Craig-Geffen-Morse[27,28]	X	X	X	X
Roberts[29]	X	X	X	X
Higgins-Leighton[30,31,32,33]	X	X	X	X
6. Douglas-Blair-Wagner[35]	X	X	X	X

* Categorized according to classification scheme of Table 3-1.
** k_{rw} only.

S_{or} = residual water saturation, fraction.
S_w = connate water saturation, fraction.
ϕ = porosity, fraction.
E_p = primary recovery efficiency, fraction of original oil-in-place.
E_o = overall sweep efficiency, fraction of reservoir volume.

This equation is based on the assumption that the unswept portion of the reservoir at the time of flood abandonment is completely saturated with oil and connate water. It is also assumed that there is no gas cap and that a free gas saturation does not exist in the swept portion of the reservoir.

Goolsby[50] suggested the following steps for flood design:

General Principles of Waterflood Design 53

Init. and resid. oil saturations	Initial gas saturation	Resid. gas saturation	Oil and water viscosity	Gas viscosity	Average injection rate	Injection history	Correlation for areal sweep
X	—	—	X	—	X	—	—
X	—	—	X	—	X	—	—
X	X	X	X	—	X	—	—
X	—	—	X	—	X	—	—
X	—	—	X	—	X	—	—
X	—	—	X	—	X	—	—
X	—	—	X	—	X	—	—
X	X	—	X	—	X	—	—
X	X	—	—	—	—	X	—
X	—	—	X	—	—	—	—
X	X	X	X	X	—	X	X
—	—	—	—	—	—	—	—
X	—	—	—	—	X	—	X
X	—	—	X	—	—	X	X
—	—	—	—	—	—	—	—
—	—	—	X	—	—	—	—
X	X	—	X	—	—	—	—
X	X	X	X	—	—	X	X
X	—	—	X	—	X	—	—
X	X	X	X	—	—	X	X
X	X	X	X	X	—	—	—

1. Characterize the reservoir geologically as completely as possible.
2. Determine the values for Callaway's[49] five "primary variables." Most of these variables can be expressed with a range developed from field and laboratory data and the study of analogous fields.
3. Calculate a range of waterflood recoveries using Eq. 3-1.
4. Relate the maximum, average, and minimum recoveries to time by using fluid-in–fluid-out and water–oil production relationships taken from analogous fields. The effects of various injection rates can be incorporated in this step if felt necessary.

The "primary variables" can also be expressed as probability distributions. Equation 3-1 is then solved using these distributions yielding a

TABLE 3-5
Comparison of Basic Methods Concerned with Permeability–Heterogeneity–Injectivity Problems[a]

	Dykstra–Parsons[1]	Stiles[5]	Yuster–Suder–Calhoun[10]
Reservoir configuration	Stratified-linear	Stratified-linear	Stratified five-spot
Permeability distribution	Normal probability	Rearranged actual data	Average of permeability capacity
Injection controlled	Mobility ratio	kh capacity	kh capacity
Total injection rate	Constant	Constant	Variable (to fill-up), constant (after fill-up)
Layer injection rate	Variable	Constant	Variable (to fill-up), constant (after fill-up)
Required mobility ratio	Any	1.0	1.0
Gas fill-up	Before oil production	Initially	By individual beds
Displacement mechanism	Pistonlike	Pistonlike	Pistonlike
Areal sweep	100% at WBT[b]	100% at WBT[b]	Sweep efficiency factor[c]
Vertical sweep	Proportional to permeability capacity and mobility ratio	Proportional to permeability capacity	Recovery factor[c]
Solution method	Graphical	Graphical and numerical	Numerical

[a] After Schoeppel,[44] courtesy of *Oil and Gas Journal*.
[b] WBT = Water breakthrough.
[c] Determined as a function of throughput.

General Principles of Waterflood Design

TABLE 3-6

Comparison of Basic Methods Concerned with Areal Sweep Efficiency Problems[a]

	Muskat[13]	Hurst[14]	Atlantic–Richfield[15–18]
Reservoir configuration	Stratified five-spot	Stratified five-spot	Stratified five-spot
Total injection rate	Constant	Constant	Variable
Applicable mobility ratio	1.0	1.0	0.1–10
Initial gas saturation	None	None	None
Sweep efficiency at WBT[b]	72.3%	72.6%	Variable
Sweep efficiency after WBT[b]	Not available	Not available	From correlation

[a] After Schoeppel,[44] courtesy of *Oil and Gas Journal*.
[b] WBT = Water breakthrough.

probability distribution for the waterflood recovery instead of a simple range-average value.

Sweep Efficiency

The effectiveness of a secondary recovery process is dependent on the volume of the reservoir which will be contacted by the injected fluid.

TABLE 3-7

Waterflood Prediction Methods Concerned with Displacement Mechanism

Basic method	Highlights of method
Buckley–Leverett (1942)[23]	Material balance on element; variable-saturation flood front displacement mechanism; single layer–linear model; constant injection rate; no residual gas saturation; 100% sweep at water breakthrough.
Modification	
1. Terwilliger et al. (1951)[24]	Saturations in flood front; stabilized zone concepts.
2. Felsenthal–Yuster (1951)[25]	Radial case
3. Welge (1952)[26]	Average saturation at water breakthrough
4. Craig–Geffen–Morse (1955)[27,28]	Five-spot pattern; sweep efficiency correlated with mobility ratio at breakthrough.
5. Roberts (1959)[29]	Allowance for stratification
6. Higgins–Leighton (1960)[30,31,32,33]	Channel and cell displacement; any well pattern.

[a] After Schoeppel,[44] courtesy of *Oil and Gas Journal*.

The latter, in turn, is dependent on the horizontal and vertical sweep efficiency of the process. The following factors control the sweep efficiency:

1. Pattern of injectors.
2. Off-pattern wells.
3. Unconfined patterns.
4. Fractures.
5. Reservoir heterogeneity.
6. Continued injection after breakthrough.
7. Mobility ratio.
8. Position of gas–oil and oil–water contacts.

Pattern Selection

The selection of an injection pattern is one of the first steps in the design of secondary recovery projects. When making the choice, it is necessary to consider all the available information about the reservoir. The adverse effects of the other factors listed above can be partially offset if they are considered during the pattern selection. Other factors which should be considered in pattern selection are as follows:

1. Flood life.
2. Well spacing.
3. Injectivity.
4. Response time.
5. Productivity.

The flood life depends on the availability of water, the rate at which water can be injected, well spacing, and proration policies. The performance and economics for various well spacings and pattern sizes should be analyzed in order to arrive at the economically optimum choice. These analyses, however, cannot be made without also considering injectivity, which is best determined from pilot operations. A well-controlled pilot operation is essential to understanding all the pattern selection factors.

Empirical methods for estimating water injectivity prior to an actual test for pattern floods have been worked out by Muskat[13] and by Deppe[21] for a mobility ratio of unity in the case of single fluid flow, and are presented in Fig. 3-1. Prats et al.[12] have developed a plot relating dimensionless injectivity, I_D, to mobility ratio, $M_{w,o}$, for different stages of a flood in five-spot patterns when the reservoir has an initial mobile gas saturation (Fig. 3-2). Dimensionless injectivity, I_D, is defined by the following equation:

$$I_D = \frac{i_w \mu_w}{k_w h \Delta p} \qquad (3\text{-}2)$$

General Principles of Waterflood Design

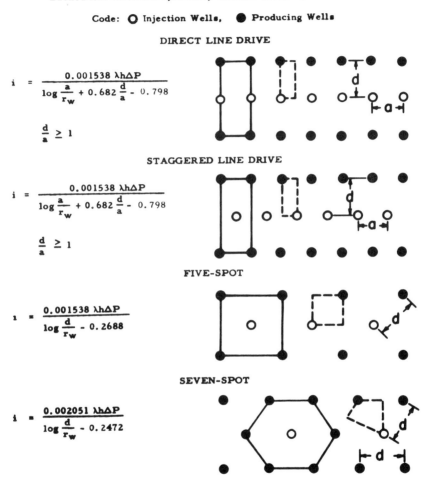

FIG. 3-1. Injectivities for regular patterns for mobility ratio equal to one. h = net pay thickness, ft; r_w = wellbore radius, ft; Δp = difference in pressure between the injection well and producer, psi; and λ = mobility of reservoir fluid, md/cp. (After Deppe,[21] courtesy of the SPE of AIME.)

NINE-SPOT

$$i = \frac{0.001538 \lambda h \Delta P_{i,c}}{\frac{1+R}{2+R}\left[\log \frac{d}{r_w} - 0.1183\right]},$$

$$i = \frac{0.003076 \lambda h \Delta P_{i,s}}{\frac{3+R}{2+R}\left[\log \frac{d}{r_w} - 0.1183\right] - \frac{0.301}{2+R}},$$

R = Ratio of producing rates of corner well (c) to side well (s),

$\Delta P_{i,c}$ = Difference in pressure between injection well and corner well (c),

$\Delta P_{i,s}$ = Difference in pressure between injection well and side well (s).

DIRECT LINE DRIVE BOUNDARY PATTERN

$$i = \frac{0.003076 \lambda h \Delta P_{i1',q1'}}{\frac{q_1 + 4q_1'}{q_1 + 2q_1'}\left[\log \frac{d}{r_w} + 0.166 + 0.400R\right] + \frac{1.368 q_1(1+R)}{q_1 + 2q_1'}},$$

$$R = \frac{q_2}{q_1} = \frac{q_2'}{q_1'},$$

$\Delta P_{i1',q1'}$ = Difference in pressure between injection well i_1' and producing well q_1'.

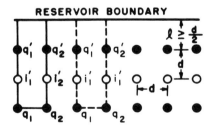

FIVE-SPOT BOUNDARY PATTERN

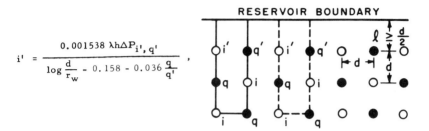

$$i' = \frac{0.001538 \, \lambda h \Delta P_{i',q'}}{\log \frac{d}{r_w} - 0.158 - 0.036 \frac{q}{q'}},$$

$\Delta P_{i',q'}$ = Difference in pressure between injection well i' and producing well q'

(Wells of outer row are alternating injection and producing wells.)

FIVE-SPOT BOUNDARY PATTERN

$$i' = \frac{0.003076 \, h\lambda \, (q_1 + q_2 + 2q_1' + 2q_2') \Delta P_{i',q_1'}}{2q_1'\left[(3+R)\{\log \frac{d}{r_w} + 0.815\} - 1.7216\right] + q_1(1+R)\{\log \frac{d}{r_w} - 0.098\}},$$

$$R = \frac{q_2}{q_1} = \frac{q_2'}{q_1'}$$

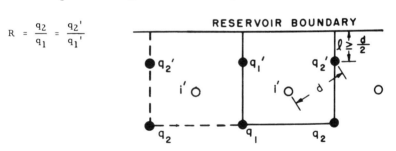

$\Delta P_{i',q'}$ = Difference in pressure between injection well i' and producing well q'.

(Wells of outer row are all producing wells.)

NINE-SPOT BOUNDARY PATTERN

$$i' = \frac{0.003076 \, \lambda h \Delta P_{i', s'}(2q_s' + 2q_c' + 3q_s + q_c)}{2q_s'\left[(2+R)\{\log\frac{d}{r_w} + 0.548\} - 1.49\right] + q_s\left[(3+R)\{\log\frac{d}{r_w} + 0.885\} - 0.577\right]},$$

$$i' = \frac{0.003076 \, \lambda h \Delta P_{i', c'}(2q_s' + 2q_c' + 3q_s + q_c)}{2q_s'\left[(1+2R)\{\log\frac{d}{r_w} + 0.204\} + 0.362\right] + q_s\left[(3+R)\{\log\frac{d}{r_w} - 0.460\} + 0.124\right]},$$

$$R = \frac{q_c}{q_s} = \frac{q_c'}{q_s'},$$

$\Delta P_{i', s'}$ = Difference in pressure between injection well i' and producing well s'.

$\Delta P_{i', c'}$ = Difference in pressure between injection well i' and producing well c'.

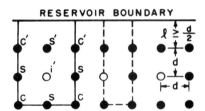

Fig. 3-1. (Continued).

where:

i_w = water injection rate, B/D.
μ_w = water viscosity, cp.
k_w = permeability to water, md.
h = thickness of injection zone, ft.
Δp = differential injection pressure, psi.

Response time is dependent on injectivity and spacing. It is further influenced by reservoir heterogeneity and the oil, gas, and water saturations which exist at the beginning of injection.

The pattern chosen must above all consider the physical characteristics of the reservoir. Formal patterns such as the 5-, 7-, or 9-spot and the direct

General Principles of Waterflood Design 61

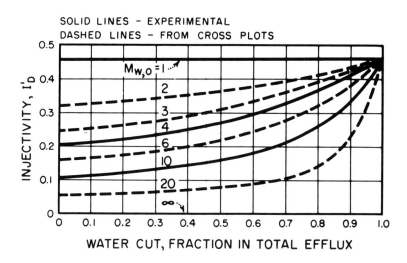

Fig. 3-2. Injectivity as a function of water cut and mobility ratio. (After Prats et al.,[12] courtesy of the SPE of AIME.)

or staggered line drives are useful only when the reservoir is generally uniform in character. Faulting and localized variations in porosity or permeability lead to irregular patterns or peripheral injection systems. For limestone reservoirs, the irregular pattern or peripheral system is more likely to be used.

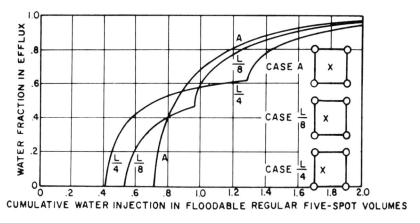

Fig. 3-3. Water-cut history of a laterally displaced production well. (After Prats et al.,[51] courtesy of the SPE of AIME.)

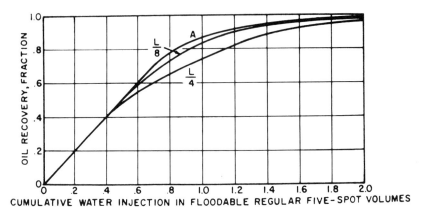

FIG. 3-4. Oil-production history of a laterally displaced production well. (After Prats et al.,[51] courtesy of the SPE of AIME.)

Off-Pattern Wells

Prats et al.[51] have calculated the effect of off-pattern wells (producers and injectors) on the performance of a regular five-spot waterflood. Data developed by their work are presented in Figs. 3-3 through 3-11. These results are strictly applicable only when the assumptions on which their work is based are met. These assumptions are (1) the reservoir is thin

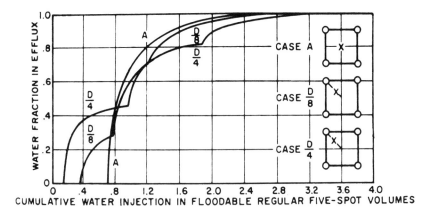

FIG. 3-5. Water-cut history of a diagonally displaced production well. (After Prats et al.,[51] courtesy of the SPE of AIME.)

General Principles of Waterflood Design

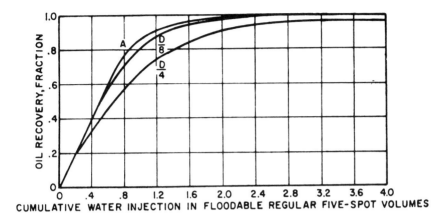

FIG. 3-6. Oil-production history of a diagonally displaced production well. (After Prats et al.,[51] courtesy of the SPE of AIME.)

and horizontal, (2) porosity, permeability, and thickness are uniform, (3) only crude oil is mobile initially in the formation, (4) mobility ratio is one, (5) injection and production rates are the same for all wells throughout the field, (6) there is a sharp boundary between the oil and water banks, (7) producing wells are kept flowing even after the individual wells have reached an economic limit cut of 98%. The data developed by Prats et al. can be used, however, to estimate the impact for other con-

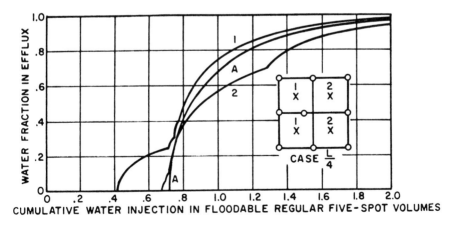

FIG. 3-7. Water-cut history of five-spot patterns surrounding laterally displaced injection well. (After Prats et al.,[51] courtesy of the SPE of AIME.)

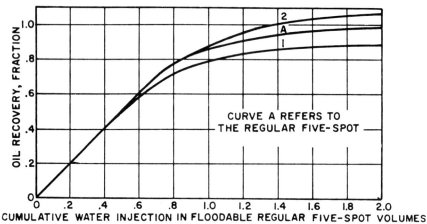

FIG. 3-8. Oil-production history of five-spot patterns surrounding laterally displaced injection well. (After Prats et al.,[51] courtesy of the SPE of AIME.)

ditions. These estimates would generally represent the minimum impact to be expected for offset wells.

Multiple irregular patterns in a single field can be evaluated with Figs. 3-3 through 3-11 as long as the irregular patterns are separated by at least one normal five-spot pattern. In Figs. 3-3 through 3-10, the A curves represent the performance of a normal five-spot flood. The other curves are keyed to the well diagrams included in the water-cut history figures. The D or L notations refer to the diagonal or lateral displacement of the wells.

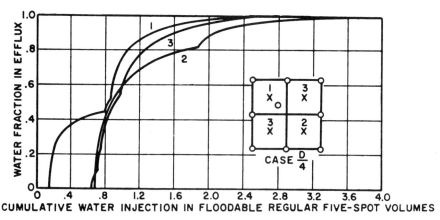

FIG. 3-9. Water-cut history of five-spot patterns surrounding diagonally displaced injection well. (After Prats et al.,[51] courtesy of the SPE of AIME.)

General Principles of Waterflood Design

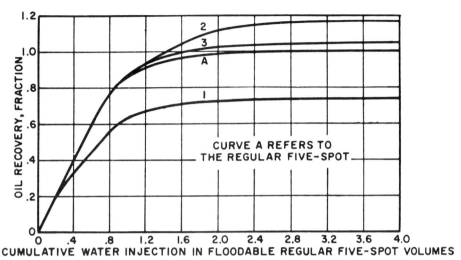

FIG. 3-10. Oil-production history of five-spot patterns surrounding diagonally displaced injection well. (After Prats et al.,[51] courtesy of the SPE of AIME.)

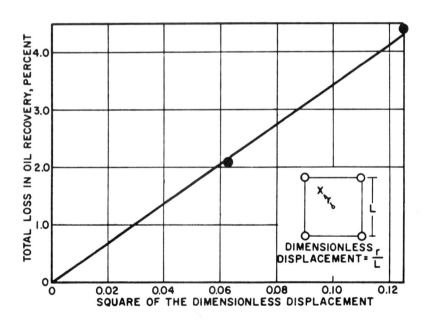

FIG. 3-11. Effect of displacement on total loss in oil recovery. (After Prats et al.,[51] courtesy of the SPE of AIME.)

Unconfined Patterns

The principal problem associated with an unconfined pattern is the loss of injected energy to wells and/or aquifer outside of the injection pattern. The degree of pattern confinement is dependent upon the reservoir pressures around this pattern. Geologic barriers such as faults and pinchouts or the presence of high-pressure aquifers limit the loss of energy. Unconfined boundaries include boundaries adjacent to (1) leases where the reservoir is unpressured, and (2) low-pressure aquifers.

Exact percentages of energy loss or oil migration outside the confined pattern will vary with overall well configuration and estimates should be made as to the magnitude of the anticipated loss. A simple procedure often used to estimate the loss is based on the assumption of uniform radial flow of the injected water from the injector. The peripheral injectors are connected by lines on a map and the losses are calculated for each injector by measuring the exterior angle between the lines connecting the injector with the injectors on either side and dividing by 360° (an injector on the side of a project may lose around 50%, whereas a corner injector may lose about 75%).

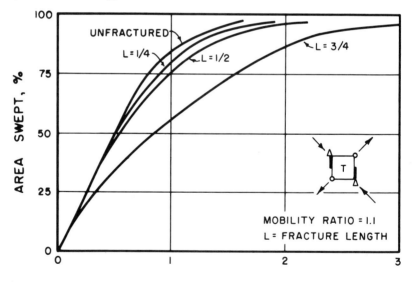

FIG. 3-12. Sweep-out with vertical fracture of favorable direction. (After Dyes et al.,[52] courtesy of the SPE of AIME.) Fracture length, L = fraction of distance between fractured well and boundary of element in flood pattern.

General Principles of Waterflood Design 67

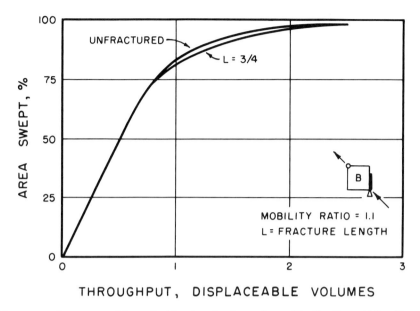

FIG. 3-13. Sweep-out with vertical fracture having unfavorable direction. (After Dyes et al.,[52] courtesy of the SPE of AIME.)

Fractures

Dyes et al.[52] studied the effect of fractures on the sweep efficiency of a five-spot pattern. Figures 3-12 through 3-15 present a summary of their findings; in the figures, the fracture length, L, is expressed as the fraction of the distance between the fractured well and the boundary of the element in the flood pattern as shown in the small diagrams located in the figure.

Figure 3-12 shows that a vertical fracture, when located in a favorable direction, has little effect on sweep efficiency. When the vertical fracture is located in an unfavorable direction, the breakthrough sweep efficiency drops with increasing fracture length, L. The ultimate sweep efficiency is unaffected if no restriction is placed on the amount of water injected or the water–oil ratio of the producer.

The work of Dyes et al.[52] was directed toward investigating the impact of fracturing techniques on waterflooding operations. Their work, however, can also be used when considering natural fracture systems. By using the trends presented, patterns can be planned to minimize the impact of the fracture system.

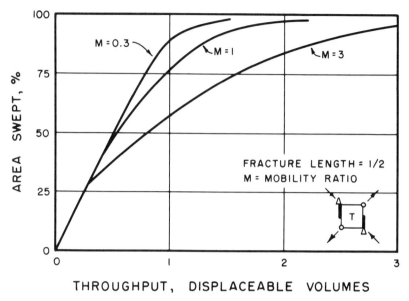

Fig. 3-14. Influence of mobility ratio and fracture on % area swept. (After Dyes et al.,[52] courtesy of the SPE of AIME.)

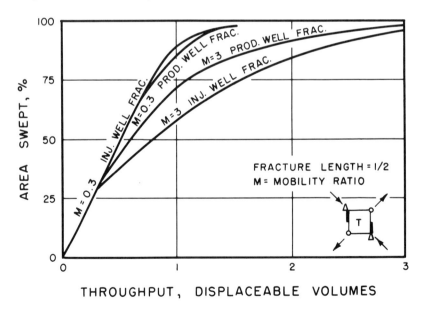

Fig. 3-15. Influence of injection or production well fracture on % area swept. (After Dyes et al.,[52] courtesy of the SPE of AIME.)

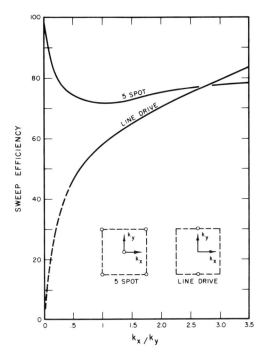

FIG. 3-16. The effect of directional permeability on sweep efficiency. (After Landrum and Crawford,[53] courtesy of the SPE of AIME.)

Reservoir Heterogeneity

Landrum and Crawford[53] have studied the effects of directional permeability on waterflood sweep efficiency and productive capacity. Figures 3-16 and 3-17 illustrate the impact of directional permeability variations on sweep efficiency for a line drive and a five-spot pattern flood, whereas Figs. 3-18 and 3-19 present the variations in productive capacity for the same conditions. These data were generated through theoretical calculations and potentiometric studies assuming a mobility ratio of one and steady state flow. Gravity and capillary effects were neglected.

The effects of directional permeability can be minimized through adjustment of the geometry of the injector–producer system. Pattern distances should be lengthened in the direction of the greater permeability.

Continued Injection after Breakthrough

Continued injection after breakthrough can result in substantial increases in recovery, especially in the case of an adverse mobility ratio. Most of the

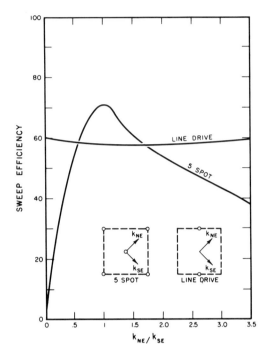

FIG. 3-17. The effect of directional permeability on sweep efficiency. (After Landrum and Crawford,[53] courtesy of the SPE of AIME.)

published data on sweep efficiency beyond breakthrough have been obtained on porous-plate or sand-pack models without an initial gas saturation. The five-spot pattern is the one most extensively studied.

The work of Craig et al.[27] has shown that the breakthrough of injected fluid at the producer is not the end of a successful flooding operation, because significant quantities of oil may be swept by water after breakthrough.[54] The higher the mobility ratio (crude gravity decreasing) the more important is the "after breakthrough" production. Figure 4-5, p. 106, shows that the breakthrough efficiency changes from a low of 51% at a mobility ratio of 10 to a high of 100% at a mobility ratio of 0.17.

Mobility Ratio

The mobility ratio is defined as the ratio of the mobility of the driving phase (such as water or LPG) to the mobility of the driven phase (such

General Principles of Waterflood Design

as oil). The mobility ratio, M, may be represented as

$$M = \frac{k_w \mu_o}{\mu_w k_o} \tag{3-3}$$

where:

k_w = water permeability, md.
k_o = oil permeability, md.
μ_w = water viscosity, cp.
μ_o = oil viscosity, cp.

The k_w refers to the effective water permeability behind the water–oil front and k_o refers to the effective oil permeability ahead of the front. This equation shows that high oil viscosities give rise to high, or unfavorable, mobility ratios (high mobility ratios generally result in poor recoveries). Figure 3-20 is a schematic drawing of the regions of a reservoir being subjected to frontal displacement with one movable phase, water,

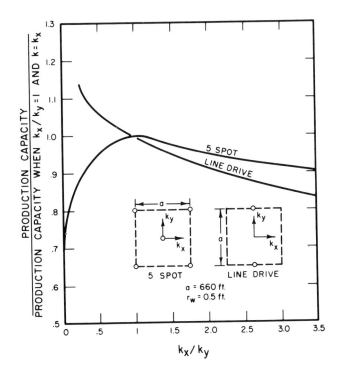

FIG. 3-18. The effect of directional permeability on production capacity. (After Landrum and Crawford,[53] courtesy of the SPE of AIME.)

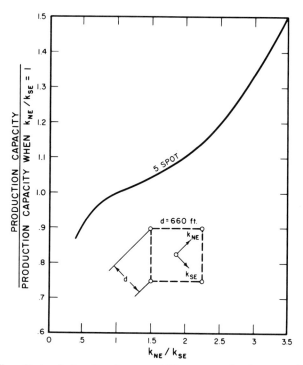

FIG. 3-19. The effect of directional permeability on production capacity. (After Landrum and Crawford,[53] courtesy of the SPE of AIME.)

behind the front and one, oil, in front. Figure 3-21 is a schematic drawing of the regions of a reservoir being subjected to partial frontal displacement (two movable phases, oil and water, exist behind the front). The effective mobility, M_e, behind the front may be represented as

$$M_e = \left(\frac{k_w}{\mu_w} + \frac{k_o}{\mu_o}\right)_2 \tag{3-4}$$

where the subscript 2 refers to the region behind the front. For partial frontal displacement, the mobility ratio, M, is given by

$$M = \left(\frac{k_w}{\mu_w} + \frac{k_o}{\mu_o}\right)_2 \bigg/ \left(\frac{k_o}{\mu_o}\right)_1 \tag{3-5}$$

The effective or relative permeability to oil and to water can be obtained from: (1) use of published data such as given by Leverett and Lewis[55]; (2) displacement tests from which k_w/k_o curves can be developed; (3) water-

General Principles of Waterflood Design

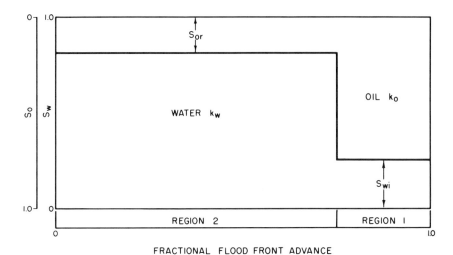

Fig. 3-20. Frontal displacement process with one movable phase behind the front.

flood tests; oil permeability is measured before a core is flooded, and water permeability is measured at the end of the test.

For reservoirs that have a moderate or large interstitial gas saturation, displacement tests and/or waterflood tests should be made in a manner that accounts for this gas-occupied pore space. Favorable mobility ratios

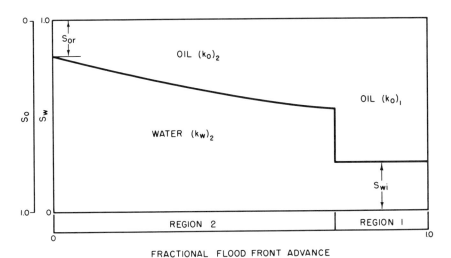

Fig. 3-21. Partial frontal displacement with two movable phases behind the front.

are equal to or are less than 1.0 and give rise to stable fronts and good recoveries. Unfavorable, or poor, mobility ratios are greater than 1.0 and have a tendency to cause unstable fronts. Recoveries become worse with increasing mobility ratios.

Major Predictive Techniques

The major predictive techniques presented here in outline form include the Buckley–Leverett, Dykstra–Parsons (Johnson), and Stiles procedures.

Buckley–Leverett Predictive Technique[23]

Two significant variations on the basic Buckley–Leverett predictive techniques have evolved: (1) the layered Buckley–Leverett and (2) the double Buckley–Leverett. These variations are treated in detail immediately following the discussion of the basic technique.

Assumptions for Basic Buckley–Leverett Method

1. A flood front exists, with only oil moving ahead of the front. Oil and water move behind the front.
2. Reservoir is a single homogeneous layer. Cross-sectional area to flow is constant.
3. Linear steady-state flow occurs and Darcy's law applies (q injected = q produced), where q is expressed in B/D.
4. There is no residual gas saturation behind the front.
5. Fractional flow of the displacing and displaced fluids after breakthrough is assumed to be a function of the mobility ratio of the two fluids (capillary and gravity effects are neglected) as expressed below:

$$f_w = \frac{1}{1 + (k_{ro}\mu_w/k_{rw}\mu_o)} \tag{3-6}$$

where:

k_{ro} = relative permeability to oil, fraction.
k_{rw} = relative permeability to water, fraction.
μ_o = oil viscosity, cp.
μ_w = water viscosity, cp.

6. Fill-up occurs in all layers prior to flood response. The flood life should be increased to reflect the fill-up period.

General Principles of Waterflood Design

TABLE 3-8
Organization of Relative Permeability Data for the Buckley–Leverett Waterflood Predictive Technique

(1)	(2)	(3)	(4)	(5)	(6)
S_w	k_{ro}	k_{rw}	k_{ro}/k_{rw}	μ_w/μ_o	f_w

Procedure for Basic Buckley–Leverett Method

1. Organize relative permeability data into form suggested in Table 3-8. If several sets of relative permeability data exist for a reservoir, use the set which is representative of the portion of the reservoir to be flooded.
2. Calculate the fractional flow, f_w, as a function of water saturation, S_w, using Eq. 3-6 and plot on cartesian coordinate paper as shown in Fig. 3-22.
3. Draw tangent to fractional flow curve as indicated in Fig. 3-22. This gives the water saturation value at the flood front at breakthrough. The average saturation behind the front is read at $f_w = 1.0$.
4. Determine graphically the rate of change in the fractional flow, f_w', as a function of the change in the floodfront water saturation.

$$f_w' = df_w/dS_w \approx \Delta f_w/\Delta S_w \qquad (3\text{-}7)$$

5. Draw 6 to 8 tangents to the fractional flow curve at S_w values greater than that at breakthrough. Determine the S_{wa} and f_w' values corresponding to these S_w points.
6. Plot f_w' versus S_w at the flood front on cartesian coordinate paper and draw a smooth curve through the points. Read smoothed f_w' points for each of the S_w points.
7. Calculate the recovery of oil, N_p, in barrels at breakthrough following the steps in Table 3-9,

$$N_p = 7758 A h \phi \left(\frac{S_{wa} - S_{wi}}{B_o} \right) \qquad (3\text{-}8)$$

where:

A = areal extent of the reservoir, acres.
h = average thickness of reservoir, ft.
ϕ = average porosity, fraction.
B_o = oil formation volume factor, bbl/STB.
S_{wa} = average water saturation, fraction.
S_{wi} = initial water saturation, fraction.

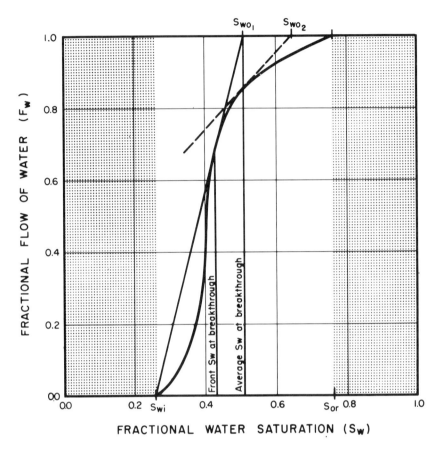

Fig. 3-22. Relationship between the fractional flow of water, f_w, and fractional water saturation, S_w.

TABLE 3-9

Recommended Steps in Using Buckley–Leverett[23] Waterflood Predictive Technique

(1)	(2)	(3)	(4)	(5)	(6)	(7)	(8)	(9)
$S_{w(\text{front})}$	f_w	S_{wa}	$S_{wa} - S_{wi}$	N_p	f_w'	WOR	W_i	t

General Principles of Waterflood Design

8. Calculate the recovery of oil to each of the S_w points using Eq. 3-8 and enter into Table 3-9.
9. Calculate the water/oil production ratio, WOR, as follows for each of the S_w points and enter into Table 3-9:

$$WOR = \frac{B_o}{(1/f_w) - 1} \qquad (3\text{-}9)$$

10. Calculate the cumulative water injected, W_i, to each of the points as follows and enter into Table 3-9:

$$W_i = \frac{7758Ah\phi}{f_w'} \qquad (3\text{-}10)$$

11. Calculate the time, t, to reach each S_w point as follows and enter into Table 3-9:

$$t = W_i/i_w \qquad (3\text{-}11)$$

If the injection rate, i_w, is not constant throughout the life of the flood, use a time weighted average rate.

12. Plot WOR versus N_p on cartesian coordinate paper. Select a WOR cutoff which is acceptable (90–98%, depending on lifting costs and other expenses).
13. Plot WOR versus time on cartesian coordinate paper. Determine the life of the flood from the WOR cutoff point.
14. Plot WOR versus W_i on cartesian coordinate paper. Determine total water injection from the WOR cutoff point.

Hovanessian[56] discussed the impact on the Buckley–Leverett procedures in the case of multiple lines of producers, with a single line of injectors. For the field studied, he concluded there was no difference in the total oil recovered or the water injected. There was an increase, however, in the life of the flood as the first line of wells slows down the rate of advance of the front to the following lines.

Buckley–Leverett Method Applied to a Layered System

Roberts[29] published a paper in 1959 describing how the Buckley–Leverett procedures could be applied to a layered system. Even with the modifications to the basic Buckley–Leverett procedure, the calculations could still be made with a desk calculator. As the value of the mobility ratio moves away from one, however, the results become more approximate. Snyder and Ramey[57] presented in 1967 a more complex method for applying the Buckley–Leverett method to a noncommunicating layered system. Their method, however, requires the use of a digital computer. A short description of their approach is presented following the discussion of

Roberts' method.[29] The following discussion is based on Roberts' method; however, certain modifications have been made along lines suggested in an Advanced Reservoir Engineering short course offered by Texas A&M University.

Assumptions for Layered Buckley–Leverett Method
1. The assumptions for the basic Buckley–Leverett technique hold for each of the layers.
2. The reservoir can be represented as a series of layers.
3. Water enters each layer in direct proportion to its capacity, kh.
4. There is no crossflow between layers.

Procedure for Layered Buckley–Leverett Method
The steps in the method described by Roberts[29] consist of calculating the performance of each layer by using the Buckley–Leverett method and then summing the recoveries from the different layers as water breaks through each layer. The first six steps of the Roberts' procedure are identical to those previously presented for the basic Buckely–Leverett technique. The remainder of the steps can be summarized as follows:

7. Segregate the core analysis data into permeability groups or layers and determine the average permeability, k, average porosity, ϕ, and average thickness, h, for each layer.
8. For each layer, calculate the capacity, kh, and percent of total capacity.
9. Calculate the injection rate, i_l, into each layer in B/D:

$$i_l = i_{total} \times (\% \text{ capacity}) \qquad (3\text{-}12)$$

10. Calculate the cumulative water injection, W_{il}, in bbl, into each layer to each S_w point as follows:

$$W_{il} = \frac{7758 A_l h_l \phi_l}{f_w'} \qquad (3\text{-}13)$$

where:

A_l = areal extent of layer l, acres.
h_l = average thickness of layer l, ft.
ϕ_l = average porosity of layer l, fraction.

11. Calculate q_{ol} and q_{wl} in B/D for each layer to each S_w point as follows:
Before breakthrough

$$q_{ol} = \frac{i_l}{B_o} \qquad (3\text{-}14)$$

$$q_{wl} = 0 \qquad (3\text{-}15)$$

General Principles of Waterflood Design

After breakthrough

$$q_{ol} = \frac{i_l}{B_o}(1 - f_w) \qquad (3\text{-}16)$$

$$q_{wl} = i_l(f_w) \qquad (3\text{-}17)$$

12. Calculate the recovery, N_{pl}, and the time, t_l, at breakthrough for each layer using the following equations:

$$N_{pl} = 7758\phi_l A_l h_l \frac{(S_{wa} - S_{wi})}{B_o} \qquad (3\text{-}18)$$

where:

S_{wa} = average water saturation at t_l, fraction.
S_{wi} = water saturation at start of flood, fraction.

Time, t_l, at breakthrough is equal to:

$$t_l = \frac{W_{il}}{i_l} \qquad (3\text{-}19)$$

13. Calculate the recovery, N_p, and the time, t_l, to each S_w point for each layer using Eqs. 3-18 and 3-19, respectively.
14. Calculate the recovery in all tracts of lower permeability at breakthrough of each layer using the following equation:

$$N_{pl_2} = \frac{(N_p)_{BTl_2}}{(W_i)_{BTl_2}}(i_{l_2}t_{l_1}) \qquad (3\text{-}20)$$

15. Plot the oil production rate for each layer versus time on a cartesian coordinate graph. Determine the total oil production rate as a function of time, and plot.
16. Plot the water production rate for each layer versus time on a cartesian coordinate graph. Determine the total water production rate as a function of time, and plot.
17. Using the total oil and water production rates, calculate the water–oil ratio as a function of time. Plot total oil, total water, and WOR versus time on a cartesian coordinate graph.
18. Determine the life of the flood by choosing an appropriate WOR cutoff point based on lifting costs and other expenses.
19. Plot the recovery from each layer as a function of time on a cartesian coordinate graph. Determine the total recovery as a function of time, and plot.
20. Determine the project's ultimate recovery using the WOR-time cutoff from step 16.

Snyder–Ramey Assumptions

The main differences between the Snyder–Ramey[57] approach and the modified Roberts' method are given here.

1. Different initial saturations, residual saturations, and relative permeability–saturation relationships can be used for each layer.
2. Each layer is divided into cells, and calculations are carried from cell to cell (this permits close observation of the movement of the front in each layer).
3. Injectivity into each layer is controlled by the resistances offered by the cells in series which vary as a function of time.

The decision to use either the modified Roberts' or the Snyder–Ramey method must be made on the basis of the quality of the input data and the time available for the analysis.

Double Buckley–Leverett Technique

The double Buckley–Leverett method of waterflood design provides for the analysis of the advance of a distinct interface behind the water–oil interface as shown in Fig. 3-23. The second interface moves at the same time as the water–oil interface, but at a different rate. The second front is associated with improved waterflooding techniques involving the use of

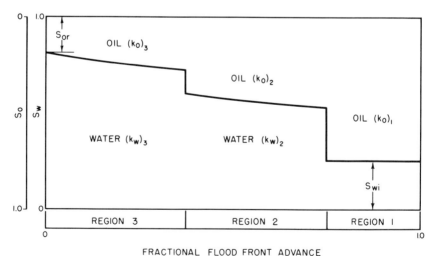

FIG. 3-23. Partial frontal displacement with two fronts and two movable phases behind each front.

General Principles of Waterflood Design

chemicals which tend to be adsorbed by the reservoir rock and alter the rock surface and/or fluid properties. The second front also occurs upon injection of hot water; in this case the heat interface travels slower than the water–oil interface. The following discussion is based on a lecture by C. E. Johnson presented at the University of Southern California in 1962.

Assumptions for Double Buckley–Leverett Method in the Case of Hot Water Injection
1. The assumptions for the basic Buckley–Leverett technique also hold for the double Buckley–Leverett method.
2. Temperature of the hot water is higher than that of the reservoir, and, as the thermal energy is transferred to the rock matrix, it is lowered to reservoir temperature.
3. Behind the temperature front (region 3 in Fig. 3-23) the reservoir fluids and rocks are heated up to the temperature of the injected hot water (allowance is made for wellbore heat losses).
4. There is no vertical heat loss from the reservoir. If included, the temperature front would advance slower.
5. The rate of advance of the temperature front, with respect to the first oil–water front, is constant dependent upon the heat capacity of the reservoir rock.
6. The increase in temperature results in an increase in the mobility of the oil and leads to more efficient displacement by the water.

Procedure for Double Buckley–Leverett Method in the Case of Hot Water Injection
1. Determine the heat (in Btu), H_g, required to heat up the reservoir rock to the temperature of the injected water:

$$H_g = V_b(1 - \phi)\rho_r C_r(\Delta T) \tag{3-21}$$

where:

V_b = bulk volume, ft^3.
$(1 - \phi)$ = fraction of bulk volume occupied by the reservoir rock.
ρ_r = density of rock, lb/ft^3.
C_r = heat capacity of rock, Btu/lb°F.
ΔT = temperature change, °F.

2. Determine the pore volume of hot water, PV_s, required to heat the reservoir rock:

$$PV_s = \frac{(1 - \phi)}{(\phi)} \frac{(\rho_r)}{(\rho_w)} \frac{(C_r)}{(C_w)} \tag{3-22}$$

where:

ϕ = porosity, fraction.
ρ_w = density of water, lb/ft³.
C_w = heat capacity of water, Btu/lb°F.

3. Determine the pore volumes of hot water, PV_p, required to heat the reservoir rock and fill the pores:

$$PV_p = PV_s + 1 \tag{3-23}$$

4. Calculate the ratio, r_v, of the velocity of the temperature front to the velocity of the water front:

$$r_v = \frac{1}{PV_p} \tag{3-24}$$

5. Solve the general Buckley–Leverett equation for the reservoir saturation at water breakthrough for a waterflood at reservoir temperature (steps 1 to 4, p. 75). The slope at this point equals the relative velocity of the cold water front at breakthrough. Calculate the volume of injected water, W_i, using Eq. 3-10.

6. Calculate the location of the heat front, in relationship to the location of the water front:

$$\frac{X}{L} = (W_i)(r_v) \tag{3-25}$$

where:

L = distance cold water has traveled.
X = distance hot water has traveled.

Note that at breakthrough of cold water, L is equal to the distance from the injector to the producer and the distance X may be solved for directly. Or, by considering volumes of fluid injected, the distance traveled by the hot front may be calculated for any volume of total fluid injected.

7. Calculate the waterflood recovery using the basic Buckley–Leverett procedures (steps 1 to 14, p. 75).
8. Correct the oil recovery to account for additional oil recovered by the heat front as follows:
   a. Assume k_w/k_o is constant and that only μ_w/μ_o changes for the temperature changes to be considered.
   b. Calculate the waterflood response using the basic Buckley–Leverett procedures (steps 1 to 14, p. 75). In step 2, use the fluid viscosities corresponding to the temperature of the hot water injected. In step 7

use a formation volume factor for hot oil (take into consideration the expansion of the crude oil due to the higher injection temperature of the water).
  c. Plot oil recovery versus water injected in the following manner:
     (1) Use cold-front recoveries until breakthrough of cold front.
     (2) Plot hot-front recoveries after breakthrough of hot front.
     (3) Show a gradual transition from cold-front recoveries to hot-front recoveries (straight line between the two breakthrough recoveries).
     (4) Select the maximum producible WOR that is acceptable and calculate the water injected to reach this WOR. Obtain the value for oil recovered from the WOR versus N_p plot.

Assumptions for Double Buckley–Leverett Method in the Case of Improved Waterfloods Using Chemicals
1. The detergent or surfactant is soluble in oil but not in water.
2. The detergent affects both the permeability and viscosity ratios.
3. A certain percentage of detergent adheres to the rock surface, whereas the remainder stays in solution.
4. The relative velocity, r_v, of the adsorbed additive to that of the water is given by

$$r_v \approx \frac{c_w}{c_w + c_a} \approx \frac{1}{1 + (c_a/c_w)} \quad (3\text{-}26)$$

where:

  c_w = concentration of additive in the water, lb/bbl.
  c_a = concentration of additive adsorbed, lb/bbl.

Procedure of Double Buckley–Leverett Method for Improved Waterfloods Using Chemicals
1. Determine the relative velocity of the additive to that of the water:

$$r_v = \frac{1}{1 + [(1-\phi)/\phi](\rho_r/\rho_w)(A/c_d)} \quad (3\text{-}27)$$

where:

A = amount of additive adsorbed per unit weight of rock, lb/lb; or lb additive/bbl water.
c_d = concentration of additive dissolved in water, lb/lb; or lb/bbl.

2. The remainder of the calculations to determine the oil recovery are similar to those discussed for hot-water floods (steps 5 through 8, p. 82.)

Dykstra–Parsons Method[1]

In the Dykstra–Parsons technique, an oil reservoir is visualized as a layered system, and recovery is calculated taking in consideration the permeability variation of this layered system and the mobility ratio.

Assumptions for Dykstra–Parsons Method
1. The reservoir consists of isolated layers of equal thickness having uniform permeability with no cross flow between layers.
2. Piston-like displacement occurs; only one phase is flowing in any given volume element.
3. There is a linear and steady-state flow.
4. The fluids are incompressible; there are no transient pressure effects.
5. The pressure drop across every layer is the same.
6. Fill-up occurs in all layers prior to flood response. The flood life should be increased to allow for the fill-up period.
7. Except for absolute permeability, the rock and fluid properties are the same for all layers.

Procedure for Dykstra–Parsons Method
The steps for calculating recovery with the aid of the coverage charts (Figs. 3-24 through 3-27) are as follows:

1. Assemble permeability data in descending order. Develop cumulative frequency distribution for the permeability values. Convert this frequency distribution to a cumulative probability distribution.
2. Plot the cumulative probability values versus log of permeability on probability paper, and draw a straight line (best fit) through the data points. Calculate permeability variation, V_k, using values from the straight line:

$$V_k = \frac{k_{50} - k_{84.1}}{k_{50}} \quad (3\text{-}28)$$

where:

k_{50} = the median permeability with 50% of the permeability values being greater than or equal to it, md.
$k_{84.1}$ = the permeability with 84.1% of the permeability values being greater than or equal to it, md.

3. Calculate mobility ratio, M:

$$M = \frac{k_w \, \mu_o}{\mu_w \, k_o} \quad (3\text{-}29)$$

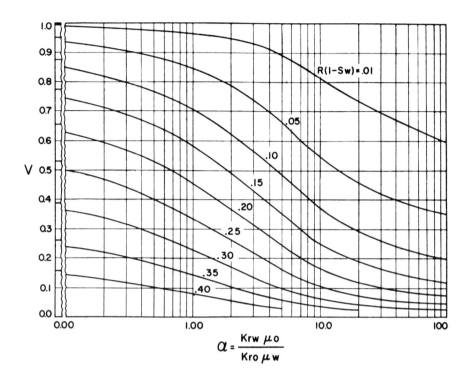

FIG. 3-24. Permeability variation (V or V_k) plotted against mobility ratio (M or α) showing lines of constant $[R(1 - S_w)]$ for a producing water-oil ratio of 1. (After Johnson,[2] courtesy of the SPE of AIME.)

4. From Johnson's[2] charts (Figs. 3-24 through 3-27) determine the coverage, R, for a WOR equal to 1, 5, 25, and 100.
5. Calculate the oil recovery, N_p, from the following equation:

$$N_p = \frac{7758 A h \phi R (S_{oi} - S_{or})}{B_o} \qquad (3\text{-}30)$$

where:

S_{oi} = initial oil saturation, fraction.
S_{or} = residual oil saturation, fraction.

6. Plot N_p versus WOR. Decide on an acceptable WOR cutoff point and read the recovery from the graph.

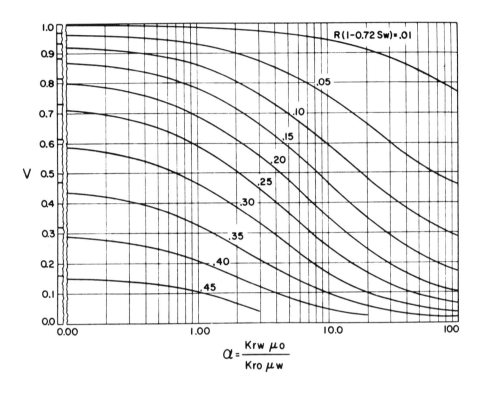

Fig. 3-25. Permeability variation (V or V_k) plotted against mobility ratio (M or α) showing lines of constant $[R(1 - 0.72 S_w)]$ for a producing water–oil ratio of 5. (After Johnson,[2] courtesy of the SPE of AIME.)

7. Integrate N_p–WOR curve graphically to get the volume of produced water, W_p.
8. Calculate the water injected, W_i:

$$W_i = N_p B_o + W_p \qquad (3\text{-}31)$$

9. Life, t, in years is given by:

$$t = \frac{W_i}{(i_w)(365)} \qquad (3\text{-}32)$$

where:

i_w = daily injection rate, B/D.

General Principles of Waterflood Design

10. Calculate oil production rate by dividing the differences in recovery by the corresponding differences in time.

Stiles Method[5]

In the Stiles method, an oil reservoir is visualized as a layered reservoir with each layer having a different permeability. Table 3-10 presents the steps in the Stiles method. The procedure steps presented here are based on a lecture by C. E. Johnson in a course on Secondary Recovery at the University of Southern California, 1962.

Assumptions for Stiles Method
1. Linear and steady-state flow occurs.
2. The reservoir is composed of isolated homogeneous layers of equal thickness with varying permeabilities; there is no crossflow between layers.
3. Except for absolute permeability, the rock and fluid properties are the same for all layers.

TABLE 3-10

Recommended Steps in Using Stiles[5] Waterflood Predictive Technique

(1)	(2)	(3)	(4)	(5)
Cumulative thickness h	Fractional thickness h_f	Permeability k_i	Dimensionless permeability K_i	Fractional permeability F_p

(6)	(7)	(8)	(9)	(10)	(11)
Capacity C_c	$(1 - C_c)$	Coverage C_e	WOR	Oil recovery N_p	Water injected W_i

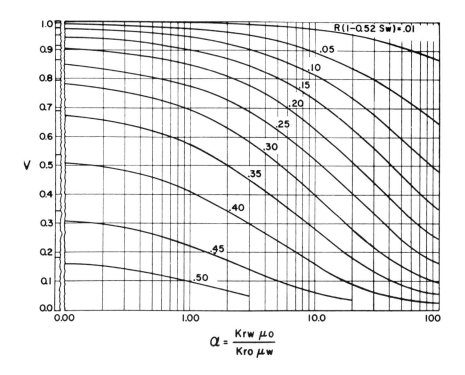

Fig. 3-26. Permeability variation (V or V_k) plotted against mobility ratio (M or α) showing lines of constant $[R(1 - 0.52\ S_w)]$ for a producing water–oil ratio of 25. (After Johnson,[2] courtesy of the SPE of the AIME.)

4. The rates of oil or water production are proportional to the quantity of water injected. All fluid movement is pistonlike.
5. Fluids are immiscible and incompressible; the pressure drop across each layer is the same.
6. The distance of flood-front penetration for each layer is proportional to its permeability.
7. Oil saturation in an unswept portion of a reservoir is unaffected by the flood front.
8. Fill-up occurs in all layers prior to flood response. The flood life should be increased to allow for the fill-up period.

Procedure for Stiles Method
1. Prepare a profile of the absolute permeabilities versus depth.
2. Divide the permeability profile into layers of equal thicknesses and select a representative permeability for each layer.

General Principles of Waterflood Design

3. Arrange the representative permeabilities in descending order with the highest value first and the lowest last.
4. Calculate the cumulative thickness, h, and fractional thickness, h_f, for the layers.
5. Determine the dimensionless permeability, K_i, for each layer and sum of all dimensionless permeabilities, K_s:

$$K_i = \frac{k_i}{k_{av}} \qquad (3\text{-}33)$$

$$K_s = \sum_{1}^{n} K_i \qquad (3\text{-}34)$$

where:

k_i = permeability to water for a layer, md.
k_{av} = average water permeability of all layers, md.

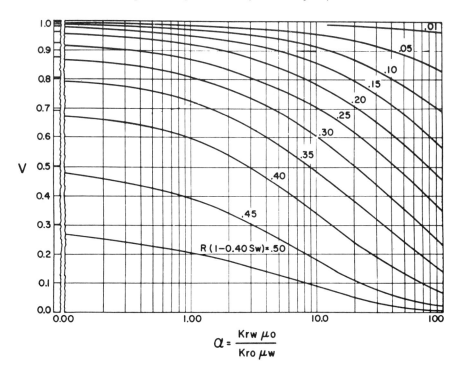

FIG. 3-27. Permeability variation (V or V_k) plotted against mobility ratio (M or α) showing lines of constant $[R(1 - 0.40\ S_w)]$ for a producing water–oil ratio of 100. (After Johnson,[2] courtesy of the SPE of AIME.)

6. Determine the incremental permeability or capacity, F_{pi}, of each layer:

$$F_{pi} = \frac{K_i}{K_s} \qquad (3\text{-}35)$$

7. Determine the cumulative capacity, C_c:

$$C_c = \sum_1^n F_{pi} \qquad (3\text{-}36)$$

8. Calculate the coverage, C_e.

$$C_e = \frac{K_i h_f + 1 - C_c}{K_i} \qquad (3\text{-}37)$$

9. Calculate the water–oil ratio, WOR:

$$\text{WOR} = \frac{C_c}{(1 - C_c)} (M)(B_o) \qquad (3\text{-}38)$$

where:

M = mobility ratio.
B_o = oil formation volume factor, bbl/STB.

10. Calculate the recovery, N_p:

$$N_p = \frac{7758 A h \phi C_e (S_{oi} - S_{or})}{B_o} \qquad (3\text{-}39)$$

where:

S_{oi} = initial oil saturation, fraction.
S_{or} = residual oil saturation (flood pot tests), fraction.

11. Plot water–oil ratio versus net oil recovery, decide upon an acceptable WOR cutoff point, and read the anticipated oil recovery from the graph.
12. Calculate the amount of cumulative water injected, W_i, and life, t, of project following steps 7 to 9, p. 86.

Improved Waterflood Processes

In the discussion of the double Buckley–Leverett predictive technique, the subject of the improved waterflood processes was mentioned briefly.

General Principles of Waterflood Design

This section provides additional information on two major groups of chemicals used to increase waterflood recovery: polymers and surfactants. The chemicals either improve the mobility ratio of the process by changing the properties of the water (polymers) or reducing the interfacial tension (surfactants).

Polymers

Polymers are used in waterflooding to reduce the mobility of the water (k_w/μ_w) and thus improve its displacement efficiency. The reduction in mobility is caused by both an increase in the water viscosity and a decrease in the permeability to water. A reduced driving phase mobility results in improvements in the areal and vertical sweep efficiencies.

During the early attempts to alter the characteristics of the water, such materials as glycerin, sugar, glycols, and some naturally occurring polymers were employed. The high concentrations of these agents required to achieve the desired changes in water properties, however, made them economically unattractive.

With the advent of the less expensive synthetic organic polymers, polymer flooding became a reality instead of a theory. Polyethylene oxides and polyacrylamides are the most common of polymers being used. The polyacrylamides are sometimes partially hydrolyzed, which further increases their molecular weight. These chemicals cause a large reduction in water mobility at low concentrations and are adsorbed only negligibly.

Mungan et al.[58] found that little reduction in residual oil saturation should be expected from polymer flooding. The increase in recovery is mainly the result of increasing the volume of the reservoir swept. Polymer flooding is best suited for reservoirs in which water sweep efficiency is very low owing to an unfavorable mobility ratio (i.e., because of low crude oil gravity) or wide permeability variations. Both conditions are common in the carbonate reservoirs.

Adsorption studies on the reservoir rock for any proposed polymer flood are absolutely essential in order to determine the polymer requirements and, therefore, the economics of the operation. Mungan et al.[58] suggested that in pilot testing a polymer flood, it would be useful to utilize a tracer ahead of the polymer slug, a tracer with the polymer, and a third tracer following the polymer. Comparison of the results obtained on using these three tracers may provide answers on the development and the transport rate of the polymer bank.

Excellent discussion of the details of polymer flooding has been presented by Pye,[59] Burcik,[60] Sandiford,[61] and Gogarty.[62]

Armstrong,[63] in reporting on polymer flooding, made the following comments:

1. A field with a large gas cap, or which is highly fractured, or has its permeability controlled by vugs is not a good candidate for polymer flooding.
2. The polymer flood is not a tertiary tool; therefore, if a waterflood is approaching the economic limit, a polymer flood should not be started.
3. Large volumes of bottom water can strip the polymer flood of chemicals.
4. Thickness of the section does not limit polymer application.
5. Depth presents no problem if the reservoir temperature is below 300°F.
6. The quality of the injected water does not affect the polymer injection.

The double Buckley–Leverett predictive technique can be used for polymer flood predictions. Laboratory work and pilot testing, however, should definitely be undertaken before a field waterflood is initiated.

A unique form of pusher flooding was used in the Crane Zone of the Northeast Hallsville Field, Harrison County, Texas.[64] The reservoir is a slightly dipping monocline with a large associated gas cap. The reservoir rock is an oolitic limestone (intergranular porosity) of varying permeability. The crude oil gravity is very high (57.1° API), and withdrawals from the gas cap were causing crude migration into the gas cap and consequent loss. To correct this, two strategic wells were chosen for polymer injection to form a viscous barrier at the gas–oil contact. After the barrier was placed, a water injection project was started in the oil zone. Water breakthrough occurred earlier than expected owing to the permeability variation. Polymer injection was tested next, and it reduced the produced water–oil ratio within 2 months from the beginning of injection. The flood has since been expanded.

Surfactants

There has been much laboratory work and some field testing of surface-active chemicals (surfactants). These tests show a reduction of the interfacial tension between the oil and water which results in a substantial reduction of the residual oil content of the reservoir rock after waterflooding.[65] Earlougher and Guerrero[66] estimated that use of surfactants might result in additional recoveries of about 10% of the original oil-in-place above that obtained by conventional waterflooding. The main cause of the failures on using surfactants is that they tend to adsorb on the surfaces of the solids; this depletes the quantity of surfactant available to work on the oil–water interface. Johnson[67] derived an equation to estimate the weight of surfactant which must be injected so that its concentration would not drop below some predetermined minimum value before reaching the end of the reservoir.

$$[(1 + kc_s)/(kc_s)]^2 = (X/L)(a/w) \tag{3-40}$$

where:

X = distance of movement of the surfactant having concentration c_s, ft.
L = total length of the reservoir, ft.
c_s = concentration of surfactant in solution in equilibrium with the adsorbed material, lb/bbl.
a = a constant, the maximum adsorption capacity per unit volume of pore space, lb/ft³.
k = the reciprocal of that solution concentration, which is in equilibrium at an adsorption capacity of $\frac{1}{2}a$, bbl/lb.
w = specific weight of surfactant, lb/ft³.

A graphic representation of the equation is shown in Fig. 3-28. This graph can be used for a radial system by changing the label on the horizontal axis to $(X/L)^2(a/w)$.[67] To use the curve, k and a must first be measured, c_m (minimum concentration effective in removing oil) must be

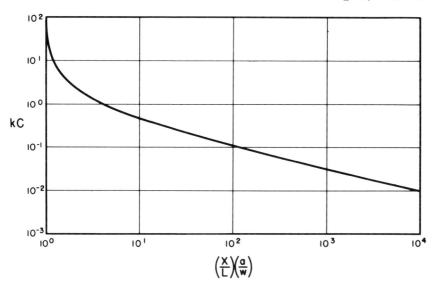

FIG. 3-28. A graphic representation of Eq. 3-40. (After Johnson,[67] courtesy of the Oil and Gas Journal.) (The writers prefer term c_s to C.)

determined experimentally, and X/L must be obtained from the physical system. The graph is entered from the left with the calculated kc_m value and the corresponding $(X/L)(a/w)$ value is read; w is then calculated.

Inks and Lahring[68] found that nonionic surfactants tend to adsorb to a lesser extent than the anionic and cationic surfactants. A field test of a nonionic surfactant yielded an apparent increase in oil recovery of about 9%.[68] Recent work by Babalyan and Kovalenko* presented an analysis of the utilization of the nonionic surfactants to increase the recovery from fissured reservoir rocks when the oil is contained mainly in the fissures or possibly large vugs. In this application, surfactants are used to lower the surface tension at the oil–water interface and the force of adhesion of the oil to the solid surface. This increases both the amount of oil recovered and the rate at which it is recovered. The actual consumption of surfactant is small, as only the fissure (fracture) walls are available for surfactant adsorption.

When the reservoir has a fracture–matrix porosity system with the majority of the oil in the matrix, or a vuggular pore system with large and small vugs, with the majority of the oil in the small vugs, it was recommended that a surfactant (anionic or cationic) be used to make the walls of the fractures or large vugs hydrophobic.* This will increase the rate of flow of the oil into fractures or large vugs.

Questions and Problems

3-1. List the six basic classifications of waterflood response prediction techniques.
3-2. For a perfect prediction technique, (a) what variables would be considered and (b) what data should be gathered?
3-3. List the types of data required for each of the following predictive techniques:
 (a) Frontal-advance (Buckley–Leverett).
 (b) Johnson's modification of Dykstra–Parsons method.
 (c) Arps' modification of the Stiles method.
3-4. For a 3,000,000 acre-ft reservoir having a porosity of 15%, determine:
 (a) Required volume of injection water in bbl.
 (b) The life of the flood assuming 3000 B/D injection rate.

---

* Babalyan, G. A. and Kovalenko, E. K.: "The Possibility of Using Surfactants to Increase the Petroleum Yield from Fissured Traps" (unpublished report). Personal communication, 1968.

General Principles of Waterflood Design

(c) Time from the beginning of injection to the first response and when the peak rate is observed.

(d) Total cost (operating expenses and capital investment) to recover the additional oil.

3-5. Utilizing Callaway's approach (p. 51), and given the following data, calculate the anticipated oil recovered from a 500,000 acre-ft reservoir:

Original formation volume factor for oil = 1.20.
Formation volume factor for oil during waterflood operations = 1.15.
Residual oil saturation = 25%.
Connate water saturation = 50%.
Initial gas saturation = 0%.
Porosity = 25%.
Primary recovery efficiency = 15%.
Overall sweep efficiency = 60%.

3-6. Determine the oil recovered by waterflood and life of the project for reservoir A, Appendix H, by use of the Buckley–Leverett method.

3-7. Utilizing the reservoir A data, Appendix H, and given the following data, calculate the oil recovery if the injection water is heated to 450°F:

Density of water = 62.2 lb/ft^3.
Specific gravity of rock = 2.68.
Heat capacity of rock = 0.3 Btu/°F.
Heat capacity of water = 1.0 Btu/°F.
Oil viscosity = 0.30 cp.
Water viscosity = 0.20 cp.
Formation volume factor = 1.15.

3-8. Utilizing the reservoir A data, Appendix H, calculate the anticipated oil recovery for a WOR of 10 using the Dykstra–Parsons technique.

3-9. Utilizing the appropriate reservoir A data, Appendix H, calculate the oil recovery by use of the Stiles method.

3-10. Why should polymers be added to injection water?

3-11. If surfactants are to be added to injection water, which types are preferred for a water-wet system?

References

1. Dykstra, H. and Parsons, R. L.: "The Prediction of Oil Recovery by Waterflood", in: *Secondary Recovery of Oil in the United States*, 2nd Ed., API (1950) 160–174.
2. Johnson, C. E., Jr.: "Prediction of Oil Recovery by Waterflood—A Simplified Graphical Treatment of the Dykstra-Parsons Method", *Trans.*, AIME (1956) **207**, 345–346.

3. Felsenthal, M., Cobb, T. R. and Heuer, G. J.: "A Comparison of Waterflood Evaluation Methods", Paper SPE 332 (Oct., 1962).
4. Felsenthal, M. and Yuster, S. T.: "A Study of the Effect of Viscosity on Oil Recovery by Waterflooding", Technical Paper presented at AIME Meeting, Los Angeles, California (Oct., 1951).
5. Stiles, W. E.: "Use of Permeability Distribution in Waterflood Calculations", Trans., AIME (1949) **186,** 9–13.
6. Schmalz, J. P. and Rahme, H. S.: "The Variation in Waterflood Performance with Variation in Permeability Profile", Producer's Monthly (1950) **14,** 9.
7. Arps, J. J.: "Estimation of Primary Oil Reserves", Trans., AIME (1956) **207,** 182–191.
8. Ache, P. S.: "Inclusion of Radial Flow in Use of Permeability Distribution in Waterflood Calculations", AIME Technical Paper 935-G (Oct., 1957).
9. Slider, H. C.: "New Method Simplifies Predicting Waterflood Performance", Pet. Eng. (1961) **33,** B68.
10. Suder, F. E. and Calhoun, J. C.: "Waterflood Calculations", Drill. and Prod. Pract., API (1949) 260–270.
11. Muskat, M.: "The Effect of Permeability Stratifications in Complete Water Drive Systems", Trans., AIME (1950) **189,** 349–358.
12. Prats, M., Matthews, C. S., Jewett, R. L. and Baker, J. D.: "Prediction of Injection Rate and Production History for Multifluid Five-Spot Floods", Trans., AIME (1959) **216,** 98–101.
13. Muskat, M.: Flow of Homogeneous Fluids, J. W. Edwards, Inc., (1946).
14. Hurst, W.: "Determination of Performance Curves in Five-Spot Waterflood", Pet. Eng. (1953) **25,** B–40.
15. Slobod, R. L. and Caudle, B. H.: "X-Ray Shadowgraph Studies of Areal Sweepout Efficiencies", Trans., AIME (1952) **195,** 265–270.
16. Dyes, A. B., Caudle, B. H. and Erickson, R. A.: "Oil Production after Breakthrough as Influenced by Mobility Ratio", J. Pet. Tech. (April, 1954) 27–32.
17. Caudle, B. H. and Witte, M. D.: "Production Potential Changes during Sweepout in a Five-Spot System", Trans., AIME (1959) **216,** 446–448.
18. Caudle, B. H., Erickson, R. A. and Slobod, R. L.: "The Encroachment of Injected Fluids Beyond the Normal Well Pattern", J. Pet. Tech. (May, 1955) 79–83.
19. Aronofsky, J. S.: "Mobility Ratio, Its Influence on Flood Patterns during Water Encroachment", Trans., AIME (1952) **195,** 15–24.
20. Aronofsky, J. S. and Ramey, H. J.: "Mobility Ratio—Its Influence on Injection and Production Histories in Five-Spot Flood", Trans., AIME (1956) **207,** 205–210.
21. Deppe, J. C.: "Injection Rates—The Effect of Mobility Ratio, Area Swept, and Pattern", Soc. Pet. Eng. J. (June, 1961) 81–91.
22. Hauber, W. C.: "Prediction of Waterflood Performance for Arbitrary Well Patterns and Mobility Ratio", Trans., AIME (1964) **231,** 95–103.
23. Buckley, S. E. and Leverett, M. C.: "Mechanism of Fluid Displacement in Sands", Trans., AIME (1942) **146,** 107–116.
24. Terwilliger, P. L., Wilsey, L. E., Hall, H. N., Bridges, P. M. and Morse, R. A.: "An Experimental and Theoretical Investigation of Gravity Drainage Performance", Trans., AIME (1951) **192,** 285.
25. Felsenthal, M. and Yuster, S. T.: "A Study of the Effect of Viscosity on Oil Recovery by Waterflooding", Technical Paper presented at AIME meeting, Los Angeles, California (Oct., 1951).

26. Welge, H. J.: "A Simplified Method for Computing Oil Recovery by Gas or Water Drive", *Trans.*, AIME (1952) **195**, 91–98.
27. Craig, F. F., Jr., Geffen, T. M. and Morse, R. A.: "Oil Recovery Performance of Pattern Gas or Water Injection Operations from Model Tests", *Trans.*, AIME (1955) **204**, 7–15.
28. Abernathy, B. F.: "Analysis of Various Waterflood Prediction Methods Using Actual Performance of Pilot Waterfloods in Carbonate Reservoirs", *J. Pet. Tech.* (Mar., 1964) 276–282.
29. Roberts, T. G.: "A Permeability Block Method of Calculating a Water Drive Recovery Factor", *Pet. Eng.* (1959) **31**, B–45.
30. Higgins, R. V. and Leighton, A. J.: "Waterflood Performance in Stratified Reservoirs", USBM RI–5618 (1960).
31. Higgins, R. V. and Leighton, A. J.: "Computer Prediction of Water Drive of Oil and Gas Mixtures Through Irregularly Bounded Porous Media–Three Phase Flow", *Trans.*, AIME (1962) **225**, 1048–1054.
32. Higgins, R. V. and Leighton, A. J.: "Waterflood Prediction of Partially Depleted Reservoirs", Technical Paper SPE 757 (Oct., 1963).
33. Higgins, R. V., Boley, D. W. and Leighton, A. J.: "Aids to Forecasting Performance of Waterfloods", *J. Pet. Tech.* (Sept., 1964) 1076–1082.
34. Hendrickson, G. E.: "History of the Welch Field San Andres Pilot Waterflood", *J. Pet. Tech.* (Aug., 1961) 745–748.
35. Douglas, J., Blair, P. M. and Wagner, R. J.: "Calculation of Linear Waterflood Behavior Including the Effects of Capillary Pressure", *Trans.*, AIME (1958) **213**, 96–102.
36. Hiatt, W. N.: "Injected-Fluid Coverage of Multiwell Reservoirs with Permeability Stratification", *Drill. and Prod. Pract.*, API (1958) 165–194.
37. Douglas, J., Peaceman, D. W. and Rachford, H. H.: "A Method for Calculating Multi-dimensional Immiscible Displacement", *Trans.*, AIME (1959) **216**, 297–308.
38. Naar, J. and Henderson, J. H.: "An Imbibition Model—Its Application to Flow Behavior and the Prediction of Oil Recovery", *Soc. Pet. Eng. J.* (June, 1961) 61–70.
39. Warren, J. E. and Cosgrove, J. J.: "Prediction of Waterflood Behavior in a Stratified System", *Soc. Pet. Eng. J.* (June, 1964) 149–157.
40. Morel-Seytoux, H. J.: "Analytical-Numerical Method in Waterflooding Predictions", *Soc. Pet. Eng. J.* (Sept., 1965) 247–258.
41. Guthrie, R. K. and Greenberger, M. H.: "The Use of Multiple-Correlation Analyses for Interpreting Petroleum-Engineering Data", *Drill. and Prod. Pract.*, API (1955) 130–137.
42. Schauer, P. E.: "Application of Empirical Data in Forecasting Waterflood Behavior", AIME Technical Paper 934 (Oct., 1957).
43. Guerrero, E. T. and Earlougher, R. C.: "Analysis and Comparison of Five Methods Used to Predict Waterflood Reserves and Performance", *Drill. and Prod. Pract.*, API (1961) 78.
44. Schoeppel, R. J.: "Waterflood Prediction Methods", *Oil and Gas J.* (Jan. 22, 1968) 72–75, (Feb. 19, 1968) 98–106, (March 18, 1968) 91–93, (April 8, 1968) 80–86, (May 6, 1968) 111–114, (June 17, 1968) 100–105, and (July 8, 1968) 71–79.
45. (Editorial): "Practical Waterflooding Shortcuts", *World Oil* (Dec., 1966) 89–92.
46. Earlougher, R. C. and Guerrero, E. T.: "New Developments in Waterflooding—

Part II, Newer Waterflooding Process and Equipment", *Producer's Monthly* (March, 1965) 9–16.
47. Bleakley, W. B.: "A Case for Total Engineering", *Oil and Gas J.* (March 15, 1965) 74–76.
48. Miller, F. H. and Perkins, A.: "Feasibility of Flooding Thin, Tight Limestones", *Pet. Eng.* (April, 1960) B-55–B-75.
49. Callaway, F. H.: "Evaluation of Waterflood Prospects", *J. Pet. Tech.* (Oct., 1959) 11–16.
50. Goolsby, J. L.: "Here's the Relation of Geology to Fluid Injection in Permian Carbonate Reservoirs, West Texas", *Oil and Gas J.* (July 31, 1967) 188–190.
51. Prats, M., Hazebroek, P. and Allen, E. E.: "Effect of Off Pattern Wells on the Performance of a Five-Spot Flood", *J. Pet. Tech.* (Feb., 1962) 173–178.
52. Dyes, A. B., Kemp, C. E. and Caudle, B. H.: "Effect of Fractures on Sweep-out Pattern", *Trans.*, AIME (1958) **213**, 245–249.
53. Landrum, B. L. and Crawford, P. B.: "Effect of Directional Permeability on Sweep Efficiency and Production Capacity", *J. Pet. Tech.* (Nov., 1960) 67–71.
54. Habermann, B.: "The Efficiency of Miscible Displacement as a Function of Mobility Ratio", *Trans.*, AIME (1960) **219**, 264–272.
55. Leverett, M. C. and Lewis, W. B.: "Steady Flow of Oil-Gas-Water Mixtures Through Unconsolidated Sands", *Trans.*, AIME (1941) **142**, 107.
56. Hovanessian, S. A.: "Waterflood Calculations for Multiple Sets of Producing Wells", *J. Pet. Tech.* (Aug., 1960) 65–68.
57. Snyder, R. W. and Ramey, H. J. Jr.: "Application of Buckley–Leverett Displacement Theory to Noncommunicating Layered Systems", *J. Pet. Tech.* (Nov., 1967) 1500–1506.
58. Mungan, N., Smith, F. W. and Thompson, J. L.: "Some Aspects of Polymer Floods", *J. Pet. Tech.* (Sept., 1966) 1143–1150.
59. Pye, D. J.: "Improved Secondary Recovery by Control of Water Mobility", *J. Pet. Tech.* (Aug., 1964) 911–916.
60. Burcik, E. J.: "What, Why and How of Polymers for Waterflooding", *Pet. Eng.* (Aug., 1968) 60–64.
61. Sandiford, B. B.: "Laboratory and Field Studies of Waterfloods Using Polymer Solutions to Increase Oil Recoveries", *J. Pet. Tech.* (Aug., 1964) 917–922.
62. Gogarty, W. B.: "Mobility Control with Polymer Solutions", *Soc. Pet. Eng. J.* (June, 1967) 161–173.
63. Armstrong, T. A.: "Polymer Floods Attacking Recovery Gap", *Oil and Gas J.* (Jan. 23, 1967) 46–48.
64. Snell, G. W. and Schurz, G. F.: "Polymer Chemicals Aid in Unique Recovery Process", *Pet. Eng.* (Feb., 1966) 53–59.
65. Andresen, K. H., Torrey, P. D. and Dickey, P. A.: "Capillary and Surface Phenomena in Secondary Recovery", in: *Secondary Recovery of Oil in the United States*, 2nd. Ed., API (1950) 233–239.
66. Earlougher, R. C. and Guerrero, E. T.: "New Development in Waterflooding, Part II, Newer Waterflooding Process and Equipment", *Producer's Monthly* (March, 1965) 9–16.
67. Johnson, C. E., Jr.: "Surfactant Slugs in Waterflooding—How Much? What Concentration?" *Oil and Gas J.* (Sept. 26, 1960) 220–226.
68. Inks, C. G. and Lahring, R. I.: "Controlled Evaluation of a Surfactant in Secondary Recovery", *J. Pet. Tech.* (Nov., 1968) 1320–1324.

CHAPTER 4

Carbonate Reservoir Waterflood Predictions and Performance

Introduction

Carbonate reservoirs can be separated into three broad categories based on their porosity systems: (1) intercrystalline–intergranular, (2) fracture–matrix, and (3) vugular–solution. Each type of porosity system presents a different set of factors which must be considered in the design of a waterflood project. Prediction methods presented in Chapter 3 may, therefore, require some modification. To increase the usability of the available reservoir performance data, they also are discussed separately for each porosity system.

Intercrystalline–Intergranular Porosity Systems

Because of the similarity in distribution and movement of fluids within sandstone and carbonate rock having intercrystalline–intergranular porosity, predictive techniques developed and used successfully for sandstone reservoirs can often be directly applied to this type of carbonate reservoir. Selection of the predictive technique depends upon the characteristics to be modeled and the available reservoir data. In a reservoir having distinct layers, the Dykstra–Parsons, Stiles, or layered Buckley–Leverett prediction technique can be used. If the mobility ratio is significantly greater than one, the Dykstra–Parsons or layered Buckley–Leverett technique is preferred. In a carbonate reservoir with a variety of permeabilities but no true layering, the Stiles method is suggested. The Buckley–Leverett technique gives excellent results in those cases where the interval under consideration has little permeability variation. If sufficient data are available and several techniques appear to be applicable, it is good practice to use at least two approaches for a comparison of predictions.

Prediction Methods and Comparison with Actual Performance

Goolsby and Anderson[1] used both the Stiles and Dykstra–Parsons techniques in evaluating a five-spot pilot waterflood in the McElroy Field, Texas. The predicted water–oil ratios were much greater than the actual ones. The core permeability variation was calculated to be near 0.85, but the available injectivity profiles showed no indication of formation stratification in the reservoir. The Buckley–Leverett technique using hypothetical relative permeability curves gave a good approximation to the actual performance. Calhoun et al.[2] also noted that efficient waterflooding is possible in a number of areas in the midcontinent despite large permeability variations within the pay zone, because the highly permeable streaks were not continuous between wells.

Henry and Moring[3] used a modification of the Stiles technique to check the performance of a pilot waterflood in the Panhandle Field, Texas. They investigated the position of the flood front in relation to the producer for a five-spot pilot. Assuming a radial disposal of the injected water and knowing the volume of water injected, the distance of water movement in each layer was calculated using the following equation:

$$W_i = [\pi\phi(1 - S_{or} - S_{gr} - S_{wi})C_f][c^2 \sum_{i=1}^{i=n} k_i^2] \quad (4\text{-}1)$$

where:

W_i = cumulative water injected, bbl.
ϕ = porosity, fraction.
S_{or} = residual oil saturation, fraction.
S_{gr} = residual gas saturation, fraction.
S_{wi} = interstitial water saturation, fraction.
C_f = conformance factor (correction factor for water lost outside flood interval, i.e., water bypassing part of reservoir).
c = proportionality constant relating the radial distance for the advance of water in a layer to the permeability of the layer.
k_i = permeability of a layer, md.
n = number of layers.

After substitution of values for S_{or}, S_{gr}, S_{wi}, C_f, and k_i, Eq. 4-1 can be reduced to

$$W_i = \text{constant} \cdot c^2 \quad (4\text{-}2)$$

Thus, knowing the amount of water injected at any point in time, c can be calculated.

The distance, r, the flood front moves in a radial system is related to the permeability of each layer by the following equation:

$$r_i = ck_i \qquad (4\text{-}3)$$

After determining c, r can be calculated for each layer. The calculated values of r can then be compared to the distance from the injector to the producer, and the number of layers which should have been watered out can be determined.

With use of the Stiles equation for the fractional flow of water, the theoretical well performance can be checked against the actual,

$$f_w = \frac{(kh)_w(B_o\mu_o k_{rw}/\mu_w k_{ro})}{(kh)_w(B_o\mu_o k_{rw}/\mu_w k_{ro}) + (kh)_o} \qquad (4\text{-}4)$$

where:

f_w = surface water cut, fraction.
$(kh)_w$ = capacity of watered-out layers, md-ft.
$(kh)_o$ = capacity of oil producing layers, md-ft.
B_o = oil formation volume factor, bbl/STB.
μ_o = oil viscosity, cp.
μ_w = water viscosity, cp.
k_{ro} = relative permeability to oil at connate water saturation.
k_{rw} = relative permeability to water at residual oil saturation.

In a review of the performance of the pilot, calculations indicated that the producer should have been producing some water. Inasmuch as this was not the case, the well was fractured to overcome suspected skin damage. The well produced 10 B/D oil and no water before the treatment and 35 B/D oil and 12 B/D water after the treatment.[3]

Abernathy[4] compared the performance of pilot waterfloods in three carbonate reservoirs, with intercrystalline–intergranular porosity, using several prediction methods: (1) Stiles,[5] (2) Craig–Geffen–Morse,[6] (3) Craig–Stiles,[4] and (4) Band.[7] The assumptions made in using these techniques are presented here.

A. *Assumptions common to all multilayered methods*
1. The formation is considered to be composed of a number of strata, continuous from well to well and insulated from crossflow between wells.
2. There is no segregation of fluids owing to gravity within any layer.
3. A high water saturation zone does not exist which could permit bypassing of the injected water.

B. *Stiles method assumptions*
1. The displacement occurs in a pistonlike manner.

2. Fill-up occurs in all layers before oil production, resulting from the flood, begins in any layer.
3. Sweep efficiency is constant after breakthrough.
4. Other than specific permeability, all layers have the same properties.

C. *Craig–Stiles method (multilayer) assumptions*
1. The information available from laboratory model studies on five-spot waterflooding applies (immiscible displacement).[6]
2. Frontal advance theory applies.
3. Each layer can have different absolute permeability, relative permeability, porosity, etc.
4. Areas ahead of the flood front are resaturated with oil.
5. After breakthrough, water injection is equal to total fluid production.

D. *Band method assumptions*
1. The information available from laboratory model flow studies on five-spot waterflooding applies (miscible displacement).[8]
2. Frontal advance theory applies.
3. Ten bands (layers) of equal pore volume but having different capacities are used.

E. *Craig–Geffen–Morse method (single-layer) assumptions*
1. The formation can be considered to be composed of a single layer.
2. The information available from laboratory model studies on five-spot waterflooding applies (immiscible displacements).[6]
3. Frontal advance theory applies.
4. Areas ahead of the flood front are resaturated with oil.
5. After breakthrough, water injection is equal to total fluid production.

A summary of the reservoir data for the three fields studied by Aber-

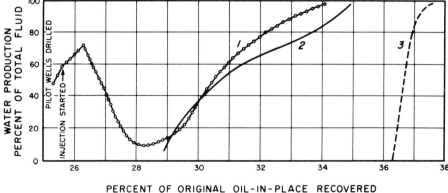

FIG. 4-1. Calculated versus actual performance of the Panhandle Field, Texas. Curve 1, actual; curve 2, calculated by Craig–Stiles technique; and curve 3, calculated by using Craig *et al.* method. (After Abernathy,[4] courtesy of the SPE of AIME.)

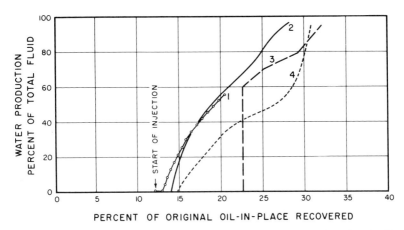

FIG. 4-2. Calculated versus actual performance of the Foster Field pilot in Texas. Curve 1, actual; curve 2, calculated by Craig–Stiles technique; curve 3, calculated by Craig et al. method; and curve 4, calculated by using Stiles method. (After Abernathy,[4] courtesy of the SPE of AIME.)

nathy[4] and a comparison of the actual recoveries with the predictions are presented in Table 4-1. Figures 4-1 to 4-3 provide a comparison of the actual results with the predictions.[4] The Craig–Stiles prediction most closely matched the actual performance for all three reservoirs.

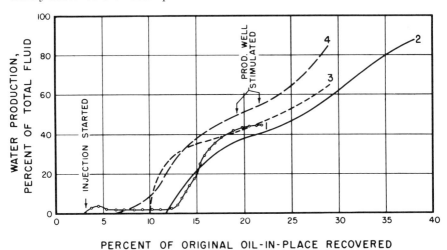

FIG. 4-3. Calculated versus actual performance of Welch Field pilot flood in Texas. Curve 1, actual; curve 2, calculated by Craig–Stiles technique; curve 3, calculated by using the Band method; and curve 4, calculated by Stiles method. (After Abernathy,[4] courtesy of the SPE of AIME.)

TABLE 4–1

Reservoir Data for Selected Carbonate Fields[a]

Field and location	Panhandle, Texas	Foster, Texas	Welch, Texas
Formation and age	Brown and White Dolomite, Permian	Grayburg–Brown Dolomite, Permian	San Andres Dolomite, Permian
Average depth, ft	3000	4200	4950
Net pay thickness, ft	65	129	75
Average porosity, %	11.7	8.6	10
Average permeability, md	7.2	2.6	6.3
Connate water saturation, %	39	23	23
FVF,[b] bbl/STB	1.044	1.061 at 244 psi	1.088 at 325 psi
Oil viscosity, cp	2.33	2.91 at 244 psi	2.32 at 325 psi
Water viscosity, cp	0.83	0.80	0.82
Stock tank oil gravity, °API	40	35	34.4
BHP[b] at start of flood, psi	—	244	325
Primary recovery at start of flood, % oil-in-place	25.6 (includes effect of gas injection)	12	2.9
Current actual recovery, % oil-in-place	34.2 (98% water cut)	21 (55% water cut)	21.7 (45% water cut)
Craig–Stiles method (layered), % oil-in-place	35 (98% water cut)	28 (98% water cut)	38 (90% water cut)
Stiles method, % oil-in-place	—	31 (98% water cut)	29 (90% water cut)
	—	28 (55% water cut)	17.5 (45% water cut)
Band method, % oil-in-place	—	—	29 (65% water cut)
	—	—	21 (45% water cut)
Craig et al. method (single layer), % oil-in-place	37.5 (98% water cut)	32.5 (98% water cut)	—

[a] After Abernathy,[4] courtesy of the SPE of AIME.
[b] FVF = Formation Volume Factor; BHP = Bottom Hole Pressure.

The steps for the Craig et al.,[6] Craig–Stiles,[4] and Band[7] methods are summarized below.

A. *Craig–Geffen–Morse method procedures*[6]
1. Follow steps 1 to 6 from the Buckley–Leverett prediction procedure (p. 75).

2. Determine k_{rw} at the average saturation behind the front at breakthrough, $(S_{wa})_{bt}$.
3. Determine k_{ro} at the oil saturation ahead of the flood front.
4. Calculate the mobility ratio, M:

$$M = \frac{k_{ru}\mu_o}{k_{ro}\mu_w} \quad (4\text{-}5)$$

5. Read the areal sweep efficiency at breakthrough, $(E_{as})_{bt}$, in percent from Fig. 4-4.
6. Determine the manner in which E_{as} changes with continued water injection using Fig. 4-5. Construct a straight line parallel to the other lines through the value $(E_{as})_{bt}$ at $W_i/(W_i)_{bt}$ equal to 1.0, where W_i = volume of cumulative water injected and $(W_i)_{bt}$ = volume of water injected to breakthrough.

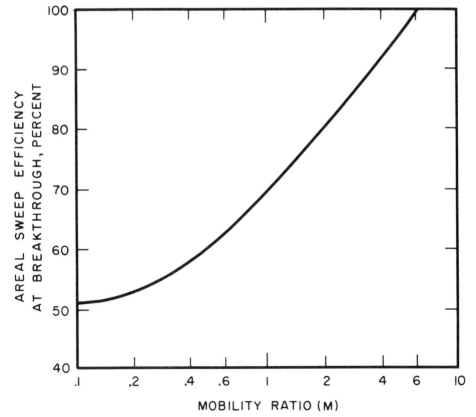

FIG. 4-4. Areal sweep efficiency at breakthrough for a five-spot well pattern. (After Craig et al.,[6] courtesy of the SPE of AIME.)

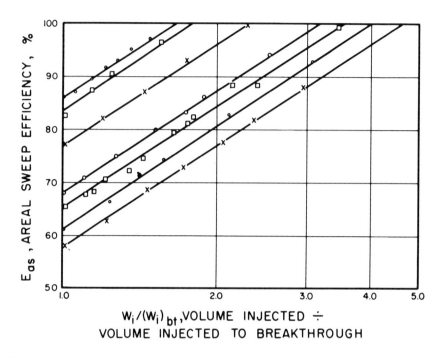

Fig. 4-5. Increase in areal sweep efficiency after breakthrough. (After Craig et al.,[6] courtesy of the SPE of AIME.)

7. Follow the steps outlined in Table 4-2. The values for the columns in the table can be determined as follows:
   (1) Select $W_i/(W_i)_{bt}$ initially as 1.00, then add maximum increments of 0.10. Smaller incrementals yield more accurate results.
   (2) $(W_i)_{bt} = [(S_{wa})_{bt} - S_{wi}](E_{as})_{bt}(PV)$ (4-6)
   where PV = pore volume of five-spot pattern, bbl; $(S_{wa})_{bt}$ = average water saturation behind flood front at breakthrough, fraction; S_{wi} = initial water saturation, fraction.
   (3) (Col. 1) × (Col. 2).
   (4) Difference between present value of W_i and that found in previous step. First value is not determined.
   (5) Read values of E_{as} for values of Col. 1 from Fig. 4-4.
   (6) (Col. 4) ÷ [(Col. 5) × (PV)]
   (7) First value = $(S_{wa})_{bt} - S_{wi}$. Subsequent values = (Col. 6) + previous value.
   (8) 1.0 ÷ (Col. 7).

(9) Develop f_w versus S_w and f_w' versus S_w curves using steps 1 to 6, p. 75 for the basic Buckley–Leverett technique. Using the value of f_w' from Col. 8, determine water saturation at the producing end of the invaded portion, S_{w2}

(10) Using the value of S_{w2} from Col. 9, determine f_w from the f_w versus S_w curve.

(11) $1.0 -$ (Col. 10).

(12) Difference between present and previous value of E_{as}. First value is not determined (blank).

TABLE 4-2

Recommended Steps in Using the Craig et al.[5] Waterflood Predictive Technique for the Five-Spot Water Injection Pattern

(1)	(2)	(3)	(4)	(5)	(6)	(7)	(8)
$W_i/(W_i)_{bt}$ ratio	$(W_i)_{bt}$ reservoir bbl	W_i reservoir bbl	ΔW_i reservoir bbl	E_{as} fraction	ΔQ_i pore volume	Q_i pore volume	f_w'

(9)	(10)	(11)	(12)	(13)	(14)	(15)	(16)
S_{w2} fraction	f_{w2} fraction	f_{o2} fraction	ΔE_{as} fraction	N_n reservoir bbl	$N_p + W_p$ reservoir bbl	N_p reservoir bbl	W_p reservoir bbl

(17)	(18)	(19)	(20)	(21)	(22)
$N_n + N_p$ reservoir bbl	Incremental oil production, STB	Average WOR over the increment, bbl/bbl	Cumulative oil recovered by waterflood, STB	Time after start of injection, days	Oil production rate after fill-up, B/D

(13) N_n = (Col. 12) × $[(S_w)_{bt} - S_{wi}](PV)$, where $(S_w)_{bt}$ = the water saturation at the front at breakthrough, fraction; and N_n = incremental oil produced from newly invaded region, bbl.

(14) (Col. 4) − (Col. 13). First value is blank. N_p = incremental oil produced from previously invaded region, bbl; and W_p = incremental water produced from previously invaded region, bbl.

(15) (Col. 11) × (Col. 14). First value is blank.

(16) (Col. 14) − (Col. 15). First value is blank.

(17) (Col. 13) + (Col. 15). First value is blank.

(18) (Col. 17) ÷ B_o. First value is blank.

(19) (Col. 16) ÷ (Col. 18). First value is blank.

(20) First value = $[((S_{wa})_{bt} - S_{wi})(E_{as})_{bt} - (S_g)_{ai}](PV)/B_o$, where $(S_g)_{ai}$ = the average initial gas saturation, fraction. Subsequent values = (Col. 18) + previous value of (Col. 20).

(21) (Col. 3) ÷ average injection rate, i_w. Plot (Col. 20) versus (Col. 21).

(22) Slope of curve of (Col. 20) versus (Col. 21).

8. Plot (Col. 20) versus (Col. 19). Choose an acceptable WOR cutoff and read the oil recovery from the curve.

B. *Craig–Stiles Method*[4] *Procedures*

Abernathy[4] presented the layered Craig–Stiles prediction method. A Craig et al. prediction[6] is made for each layer separately, and the results are superimposed to generate the overall reservoir response. The layers are chosen following the Stiles[5] system. In order to determine the relative injection rate into each layer as a function of time, the five-spot conductance ratio data developed by Caudle and Witte[9] were used (Fig. 4-6). The conductance ratio, γ, is the ratio of the water injection rate to the injection rate calculated by Muskat's single fluid five-spot equation[10] presented below (using the permeability to oil and oil viscosity):

$$ibase = \frac{3.541 \, hkk_{ro}\Delta p}{\mu_o[\ln(d/r_w) - 0.619]} \qquad (4\text{-}7)$$

where:

$ibase$ = oil injection rate, bbl/day.
h = thickness of layer, ft.
k = absolute permeability, md.
k_{ro} = relative permeability to oil at initial conditions of saturation.
Δp = pressure drop from the injection rock face to the producer rock face, psi.
μ_o = viscosity of oil, cp.
d = distance between injector and producer, ft.
r_w = wellbore radius, ft.

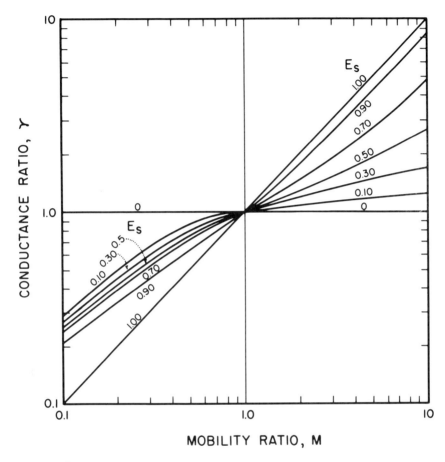

Fig. 4-6. Conductance ratio as a function of mobility ratio and the area swept, E_s (or E_{as}), for a five-spot well pattern. (After Caudle and Witte,[9] courtesy of the SPE of AIME.)

The time necessary to inject a given volume of water can be determined for each layer independently. The results at specific time intervals are determined by summing the volumes of injected water for all the layers. This procedure replaces steps 21 and 22 in Table 4-2. The following steps should be used for each layer to replace step 21.

1. Using the calculated mobility ratio (Eq. 4-5), determine the conductance ratio, γ, as a function of E_{as} using Fig. 4-6.
2. Convert the conductance ratio, γ, to a water injection rate, i_w, using

the following equation:

$$i_w = \gamma(ibase) \qquad (4\text{-}8)$$

This relationship holds when it is assumed that the pressure drop from the injector rock face to the producer rock face is constant during the life of the flood.

3. Plot i_w versus W_i. Determine the average injection rate graphically.
4. Determine time required using data from (Col. 4) of the Craig et al.[6] procedure. First value is determined from $(W_i)_{bt}$ and initial i_w.

C. *Band Method*[7]

The Band method is a modification of the Craig et al.[6] technique proposed by Hendrickson.[7] Ten bands or layers of equal volume are used in the analysis. The performance of each band is calculated, and the reservoir performance is determined by superimposition of the band results. The work of Habermann[8] is used to estimate the performance after breakthrough within each band. Inasmuch as the use of the Craig–Stiles method[4] results in better agreement with the actual performance data than the use of the Band method,[7] the details of the latter method are not presented here.

Park[11] used a combination of predictive techniques to evaluate the performance of the North Virden Scallion Field in Manitoba, Canada. Welge's method (simplification of the Buckley–Leverett technique, see p. 131) was used by Park to calculate displacement efficiencies versus water–oil ratios. The Dykstra–Parsons technique was used to determine the permeability variation, median permeability, and the coverage or vertical sweep efficiency versus water–oil ratio. The graphs of Caudle et al.[12] and Dyes et al.[13] were used to determine the areal sweep efficiency as a function of the mobility and water–oil ratios. The mobility ratio used by Caudle et al. and Dyes et al. is the reciprocal of the Dykstra–Parsons mobility ratio. The three efficiencies (displacement, vertical, and areal) were combined to obtain the recoveries at different water–oil ratios. The recovery to breakthrough was calculated to be 14% of the oil-in-place at the start of the flood. The ultimate recovery was calculated to be 28.4%, as compared to the primary ultimate recovery of 12.8%. When reported in 1965,[11] 3 years after the prediction, the actual performance was in close accord with that predicted.

Bleakley[14] reported on the evaluation system used in the Pennel and Lockout Butte Fields in Montana. The system featured computer generation of reservoir maps, calculation of reservoir volumes, and investigation of flooding patterns. The computer techniques were applicable because good core and log data were available to adequately describe the reservoirs. Data gathered during the operation of the South Pine waterflood, Montana, were also used to estimate injectivity and response time.

The model used considered the mobility ratio, vertical and horizontal displacement efficiencies, and piston-type displacement for each layer. Six layers were used. Five layers contained equal volumes of movable oil and had equal permeability capacity, kh, whereas the sixth layer contained only water to simulate the initial water cut and injected water lost to other zones.

In areas of the reservoir having good permeability, the recovery was calculated at 65 to 75% using a nine-spot pattern after the injection of 1.5 to 2.5 displaceable pore volumes. This was in close agreement with the 75% recovery observed in water-drive reservoirs and pilots on other portions of the Cedar Creek Anticline.

A production schedule for a 30-year period was determined for each reservoir. Injectivity decline owing to increased reservoir pressure was predicted for the South Pine flood. On the basis of the studies, the waterflood was started without the use of a pilot.

Waterflood Performance

The McElroy Field, Texas, was pilot tested four times before full-scale flooding was started. The large number of pilots were necessary to evaluate the wide variations in reservoir characteristics and the applicability of flooding techniques. Pilot 1 was a single injection test in the area of the reservoir having good permeability. Good injectivity and banking of oil were demonstrated by this pilot test. Pilot 2 was also in the high-permeability area of the field. It was initiated to assess the susceptibility of the reservoir to pattern flooding; an irregular five-spot pattern was used. The maximum rates and recoveries were obtained at the center well (four-way push). The performance of wells subjected to a two-way push was less efficient; and the wells influenced by only one injector, one-way push, had least efficient performance.

Pilot 3 was conducted to test the susceptibility of the tight flank areas to waterflooding. Injection rates were acceptable at injection pressures of 900 to 950 psig, and the production rate was increased. Pilot 4 was started to investigate the applicability of wide-spaced patterns (160-acre). Pilot 2 used 20-acre spacing and Pilot 3 used 40-acre spacing. The results were not available when the article was published in 1964.

Several additional operational techniques have been suggested for various carbonate reservoirs. To evaluate the performance of several injectors, tracers such as ammonium thiocyanate were added to injection water to identify which well was the source of injection water when breakthrough occurs.[11] The injector is identified by the presence or absence of the tracer. This technique is equally good for a pattern or line-drive flood.

Waterflooding in reservoirs with a primary gas cap can be undertaken as long as the loss of crude oil to the gas cap is minimized. In high-relief reservoirs, crestal gas injection in combination with peripheral water injection is the logical choice. In reservoirs having low relief, two alternatives are possible. (1) Crestal gas injection plus peripheral water injection as used in the Haynesville Field, Louisiana–Arkansas.[15] (2) Water can be injected at the gas–oil contact (in low-dip reservoirs—2° or less) along with peripheral injection. This technique has been used in the Rangely Field, Colorado[16] and in the Sholem Alechem Fault Block A Unit, Oklahoma.[17] Wilson[18] presented a method for analyzing the applicability of down-dip water displacement in the presence of a gas cap. The method requires a modification of the Dietz flow equations to handle the three mobile phases (oil, gas, and water). The criteria for stable flow in the presence of a gas cap are: (1) oil mobility, k_o/μ_o, must be greater than water mobility, k_w/μ_w, and (2) a flow rate must be greater than the critical velocities, v_c, for both the water–oil and water–gas displacements, but less than the critical velocity for the gas–oil displacement. The critical velocity may be calculated for each flowing phase from the following equation:

$$v_c = \frac{g(\rho_1 - \rho_2)\sin\alpha}{(\mu_1/k_1) - (\mu_2/k_2)} \qquad (4\text{-}9)$$

where:

v_c = critical velocity, ft/day.
g = gravitational constant, 32.174 ft/sec².
α = dip angle of formation, degrees.
ρ_1 = density of displaced fluid, lb/ft³.
ρ_2 = density of displacing fluid, lb/ft³.
μ_1 = viscosity of displaced fluid, cp.
μ_2 = viscosity of displacing fluid, cp.
k_1 = permeability of displaced fluid, md.
k_2 = permeability of displacing fluid, md.

If water is injected into the up-structure region of a dipping reservoir containing a gas cap and the water–oil mobility ratio is less than one (oil more mobile than water), four distinct types of flow can occur[18]:

Type 1. At rates of flow ranging from zero to the critical velocity for the displacement of oil by water, fingering of water occurs along the bottom of the formation; the gas does not flow from the gas cap and oil recovery is low.

Type 2. At rates of flow lying between the critical velocities for the displacement of oil by water and gas by water, the gas does not flow from the gas cap and oil is displaced from under the gas in a uniform manner.

Type 3. At rates of flow lying between the critical velocities for the displacement of gas by water and oil by gas, both oil and gas are displaced in a uniform manner.

Type 4. At rates of flow greater than the critical rate for displacement of oil by gas, gas fingering occurs along the top of the formation faster than the stable displacement of oil by water.

A summary of waterflood performance in carbonate reservoirs with intercrystalline–intergranular porosity systems is presented in Table 4-3.

Fracture–Matrix Porosity Systems

Reservoirs with a fracture–matrix porosity system differ from those having intercrystalline–intergranular porosity in that the double porosity system strongly influences the movement of fluids. The double porosity can be the result of fractures, joints, and/or solution channels within the reservoir. The pores in the matrix are not highly interconnected and porosity cannot be correlated with permeability. Joints or fissures occurring in massive formations are commonly vertical and are attributed to relief of the tensile forces during faulting or folding. The dual porosity system gives rise to a complex reservoir performance that cannot be directly modeled by conventional predictive techniques. Prediction by analogy is also difficult. In reservoirs which are fractured or have directional permeability, recovery is dependent upon injector-producer location. Therefore, unless the two reservoirs are similarly developed and possess similar lithology variations, results obtained by using analogies should only be considered as general approximations.

A review of the literature shows no predictive techniques developed specifically for waterflooding reservoirs having a fracture–matrix porosity system. Computer modeling shows the most promise for performance prediction in these complex reservoirs, but these models require a considerable amount of reservoir data to be of any value. The most reliable method for the flood design and performance prediction at present is the pilot flood test.

In designing a pilot flood, however, extreme care must be taken that it is located in a part of the reservoir that is representative of the total field. If the reservoir characteristics vary in different parts of the field, it may be necessary to undertake more than one pilot test.

The anisotropic permeability and the orientation of principal axes of the fracture systems can be evaluated by multiwell interference tests or tracer tests. The analysis of pressure buildup or falloff data can help in determining the apparent permeability, completion damage, and, possibly, the

TABLE 4-3

Summary of Waterflood Performance in Carbonate Reservoirs with Intercrystalline–Intergranular Porosity Systems

Field and state	Ref.	Zone and age	Flood type	Recovery Prim.	Recovery Sec.	Status	Special problems handled
West Lisbon, Louisiana	19	Pettit A, Cretaceous	Irregular pattern	14%	18% incr.	Just starting when updated in 1960.	Extremely thin (8 ft max.) reservoir. Irregular and streaky permeability.
Haynesville, Louisiana–Arkansas	15	Pettit A and B, Cretaceous	Peripheral water and crestal gas	18%	16% incr.	Under flood for 4 yr when reported in 1951.	Thin tight reservoir. Multiple zone with varying degree of isolation.
North Foster Unit, Texas	20	Grayburg–San Andres, Permian	Peripheral	—	—	Under flood for 3 yr when reported in 1966.	Limited basic data for flood evaluation.
Goldsmith–Cummins, Texas	21	Grayburg–San Andres, Permian	Peripheral	—	—	Under flood for 3 yr when reported in 1967.	Low permeability
Welch, Texas	7,4	San Andres, Permian	5-spot	—	38% total	Under flood for 5 yr when reported in 1961. Pilot successful.	
Umm Farud, Libya	22	Dahra B and Bu Charma, Ordovician	Peripheral	—	21–26% total	Under flood for 2 yr when reported in 1969.	Performance matched vs 4-layer, 2D-2 phase simulator.

Snyder, Texas	23	San Angelo, Permian	Irregular pattern	—	—	Under flood for 6 yr when reported in 1969. Secondary rec. running close to 95% of primary recovery.	Cooperative flood, not unitized. Reservoir sensitivity to quality of injected water found very low.
New Hope, Texas	24, 25	Bacon, Cretaceous	Peripheral	170 bbl/acre-ft	153 bbl/acre-ft	Under flood for 8 yr when reported in 1954.	Several zones of varying character.
South Cowden, Texas	26	Grayburg, Permian	5-spot	—	—	Under flood for 10 yr when reported in 1965; considered successful.	Directional permeability. Tight reservoir.
Elk Basin, Wyoming	27	Madison, Mississippian	Peripheral	—	—	Under flood for 7 yr when reported in 1969.	Complex, heterogeneous, multi-zone, reservoir. Zonal separation of drive mechanisms.
Waddell, Texas	28	San Andres, Permian	Peripheral bottom water	21%	21%	Under flood for 6 yr when reported in 1965; considered successful.	Aquifer underlying entire field.

nature of the primary and secondary porosity systems. These tests must, however, be used with caution, because it is possible that, with extremely low matrix porosity, the period of pressure buildup may not be long enough to obtain a true indication of the matrix pressure.[29] A good example of this type of performance is found in the West Edmond Field, Texas.[29]

The West Edmond Field is significantly different from most oil fields because its very tight matrix is substantially saturated with free gas. This was evidenced by (1) a lack of oil staining in the less permeable rocks, (2) gas evolution from cores which were not oil stained, (3) initial production of free gas with condensate from a mid-structure well, and (4) a lack of agreement between material balance and volumetric estimates of oil-in-place.[29]

A waterflood was initiated in the Bois d'Arc Zone of the West Edmond Field in the middle of 1949 and continued to the end of 1953, when it was shut down because of a failure to increase the oil production. When the water injection was stopped, the oil production in the general area of the flood steadily increased from 700 B/D to about 1500 B/D in 15 months and then leveled off at 900 B/D to 1000 B/D for 4 years (Fig. 4-7). One

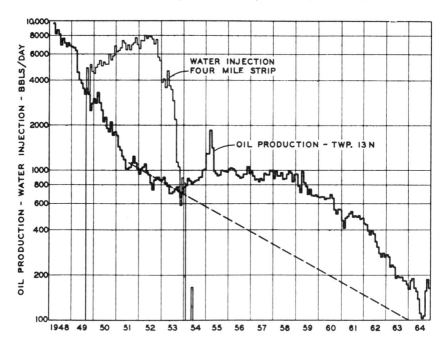

FIG. 4-7. Bois d'Arc waterflood performance, West Edmond Field, Township 13N, Texas. (After Elkins,[29] courtesy of the SPE of AIME.)

former injector was put on production in 1956. By January of 1968, 131,500 bbl of oil were recovered from the reservoir by this well, which received more than one million bbl of water during injection. This reaction was thought to be the result of a combination of imbibition flooding, pressure-pulsing, and pressure control of the flow from the matrix to the fractures (An explanation of these methods of waterflooding follows).[29] Repetition of the cycle, however, was not considered economic.

When considering the application of waterflooding or any other type of secondary recovery process, the importance of having a complete understanding of the reservoir with a fracture–matrix porosity system cannot be overemphasized. Willingham and McCaleb[30] investigated the failure of both a gas and a water injection project in the Cottonwood Creek Field in Wyoming. Both failures were attributed to the existence of an extensive fracture system, which caused the rapid channeling of the injected fluids. The fracture system was delineated through the analysis of cores, injected fluid movement, bottom-hole pressure trends, and well performance. The properties of the rock matrix were determined from lithologic studies, log analysis, and production data. The detailed analysis provided guidance as to how the operating practices had to be revised. One major operating change involved starting a pressure-pulsing program of water injection. Initial results indicated that pressure pulsing will increase oil production rates and improve ultimate recovery.

Imbibition Flooding

The low primary oil recoveries experienced in reservoirs with a fracture–matrix porosity system led operators to investigate secondary recovery techniques. As some of the world's largest reservoirs have fracture–matrix porosity systems, the development of a new method was imperative. Examples of fields with a double porosity system, created by the high degree of fracturing, are the Spraberry Field in Texas, the Kirkuk Field in Iraq, the Dukhan Field in Qatar, and the Masjid-i-Sulaiman and Haft-Gel Fields in Iran.[31] Conventional methods of fluid injection, such as waterflooding, gas injection, or miscible flooding have limited applicability to this type of reservoir because of severe bypassing of reservoir fluids by the injected fluids. In a typical formation with a fracture–matrix porosity system, over 90% of the oil is stored in the matrix blocks. On the other hand, the permeability of such a formation is mainly due to the presence of the fracture system. In a typical case, the permeability of the composite fracture–matrix system is more than 100 times that of the matrix alone.[31] Stratified reservoirs or those having a fracture–matrix porosity system are difficult, if not impossible, to flood because of bypassing. The tendency of

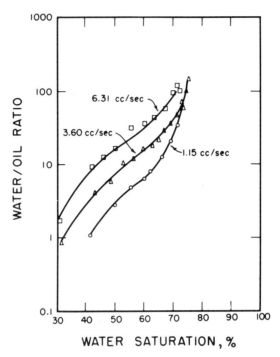

FIG. 4-8. Effect of injection rate on produced water–oil ratio for a constant fracture size. (After Graham and Richardson,[31] courtesy of the SPE of AIME.)

the injected fluid to bypass or channel through the more permeable zones is offset by the tendency of water to imbibe into the tighter zones. Imbibition may be defined as the spontaneous taking up of a liquid by a porous media.[31] Common examples of this phenomenon are dry bricks soaking up water and expelling air, a blotter soaking up ink and expelling air, and reservoir rock (preferentially water wet) soaking up water and expelling oil.

The two production mechanisms involved in recovery of oil from the matrix portion of the reservoir are displacement of oil by water flowing under (1) applied pressure gradient and (2) capillary pressure gradients.[31] At high flooding rates, the applied pressure gradients tend to control the displacement process, whereas at very low injection rates, the capillary pressure gradients dominate. When the applied pressure gradient controls the displacement process, the injected water moves along the fracture system. Very little water imbibes into the matrix, and the bulk of the oil is bypassed. When the capillary pressure gradients dominate, water imbibes readily into the matrix and displaces oil to the fracture system by both displacement and countercurrent flow.

Graham and Richardson[31] showed, through laboratory investigations, that the oil production varies directly with the oil–water interfacial tension and the square root of permeability. The rate of imbibition was found to be a complex function of the relative permeability and capillary pressure characteristics of the porous media. The presence of free gas saturation was found to decrease the rate of imbibition. Figures 4-8 and 4-9 show the effects of injection rate and the fracture–matrix permeability ratio on the water–oil ratio during flood.

Important variables in the economically successful use of imbibition are (1) the fracture spacing, which must be sufficiently close so that the area provided for water imbibing at the natural rate is large enough to give an economical rate of production from the reservoir, and (2) the displacement efficiency by natural imbibition, which must be high enough to give a desirable ultimate oil recovery.[32]

Three variations of imbibition flooding have been tested in the laboratory and in the field (primarily in the Spraberry Field, Texas). In the first variation, *low-rate injection*, part of the wells are converted to injec-

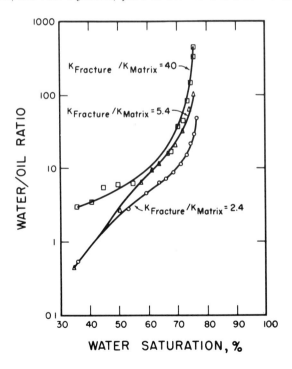

FIG. 4-9. Effect of fracture width on produced water–oil ratio for a constant injection rate. (After Graham and Richardson,[31] courtesy of the SPE of AIME.)

tors. The injection rate is kept low enough to permit the capillary forces to predominate; the water is slowly and cautiously injected. In the second variation, *cyclic injection*, part of the wells are converted to injectors, and water is injected at high rates for several months and then stopped. The water production rates of the producers drop and the oil rates increase owing to the capillary forces which are acting during the period when injection is suspended. The cycle is repeated as long as economical oil rates are obtained. In the third variation, *pressure-pulsing*, all the wells are used as injectors and producers. All wells are first used as high-rate injectors for several months. They are then shut in for 2 to 6 months before being placed on production. The cycle is repeated as long as economical oil rates are obtained.

Low-Rate Injection

A pilot test of the imbibition process was conducted in the Spraberry Field by Atlantic Refining Company using low injection rates during the 1952–55 period. It showed the process to be technically feasible but economically unsuccessful owing to low production rates. The second pilot waterflood in the Spraberry area was undertaken by Humble Oil and Refining Company in 1955.[33] A single 80-acre, five-spot pattern and an injection rate of around 500 bbl/day/well were used. The center well produced 151,000 bbl of oil by May 1962. Two former injection wells were returned to production after the injection of about 1 million bbl of water each. They subsequently each produced about 30,000 bbl of oil. The pilot test also indicated a preferential movement of water parallel to the major fault trend.[33] This preferential water movement was also observed in the pilot test conducted by the Atlantic Refining Company.

A major conclusion reached on the basis of the pilot and computer simulation studies was that a relatively high sweep efficiency can be achieved with pattern flooding if operations are conducted in a manner which takes advantage of the anisotropic permeability conditions.[33] It is necessary, however, that the producing wells be located on an axis midway between and parallel to the rows of injection wells which are located on the fracture trends. The injected water moves preferentially toward the adjacent injectors, and mutual interference occurs in a short time. Pressure buildup occurs after the fill-up of the fractures, and, consequently, the water must move in the direction perpendicular to the major fracture trends along a broad front.

The next flood started in the Spraberry area was initiated by Mobil Oil Company in 1959.[34] As in the case of Humble's test, only the upper zone was flooded. Mobil's choice of injector locations was based on the fracture orientation. The injection rates were, for most of the test, around 400 B/D.

Until October 1966, a "dump" flood operation (gravity injection) was used. In October, the injectors were equipped for pressure injection. By January 1967, the incremental oil production owing to fluid injection was equal to 83% of the primary ultimate recovery. The ultimate secondary recovery was expected to exceed the primary by the time of reaching the flood economic limit.[34]

Cyclic Injection

The next pilot flood was started by Sohio in April 1961.[35] The injectors were placed on the fault trend in the same manner as in the preceding pilots. Early breakthrough with no oil bank ahead of the water caused a reassessment of the proposed flood. A study of the pilot response and the properties of the Spraberry area lead to the conclusion that too high an injection rate had been used and that the capillary forces were not effective. Sohio engineers postulated that stopping injection might permit the capillary forces to become dominant, and that the expansive force of the fluids and rock under pressure would act to squeeze the oil out of the rock into fractures from which it can be produced.[35] This proved to be correct, and "cyclic waterflooding" was born. With the cyclic process, the performance is still dependent on capillary forces, but water is actually squeezed into the rock during the injection or pressuring-up part of the cycle. Capillary forces hold the water in the rock while the expansion forces are pushing the oil out (assuming the rock is water wet).

An evaluation of the cyclic injection process leads to the conclusion that injection equipment investment could be reduced substantially by using the same equipment in different parts of the field.[35] As the injection cycle was generally 2 months long, while the duration of the production cycle was close to 6 months, one set of injection equipment could handle three flood patterns sequentially.

In 1968, a comparison study of the Mobil and the Sohio floods was made by Sohio engineers.[36] They concluded that the long-term performances of the two floods were comparable and that lower investment and fewer operating problems were involved with the slow and steady-rate waterflooding. The oil was recovered faster with cyclic injection, but at a higher cost. Another conclusion reached was that the rate of imbibition of water into the matrix is completely inflexible and cannot be increased.

Pressure-Pulsing

In 1966, a new process, "pressure-pulsing" was initiated by Pan American Petroleum Corporation.[37] The pulsing approach appears to have several advantages over the cyclic type flooding. First, all wells may be used for water injection which should hasten fill-up and develop a more rapid and uniform increase in reservoir pressure. Second, because of increased injection pressures and rates, flooding gradients will force water into the reservoir matrix more rapidly than by imbibition alone. Third, during the pressure depletion cycle of the flood, the increased compression of the reservoir fluids and gas caused by repressuring provides energy for the displacement of the oil at a greater rate than by cyclic flooding. Fourth, during the pressure depletion cycle, all wells are used as producers resulting in a higher oil withdrawal rate than is possible with the cyclic process.[38]

Felsenthal, in laboratory tests on limestone and sandstone fractured blocks, noted that either conservation of reservoir gas or supplementing reservoir energy by gas injection appear to be necessary for the success of pressure-pulsing.[38] Energy is needed to expel the oil out while capillary action holds the water in (this is similar to the use of a gas with acid stimulation). In the laboratory tests, primary recovery was 20% of the tank oil initially in place. The first pressure-pulse cycle produced an additional 15% of the oil initially in place, but the gain in oil from pressure-pulsing dropped sharply on the following cycles.

There are also similarities between pressure-pulsing and cyclic steam injection: both need energy for oil expulsion, a soak period appears to be beneficial, and the oil recovery declines with each succeeding cycle (less oil and energy available with each succeeding cycle). This decline was also observed in the case of cyclic waterflooding.

Pressure-pulsing pilot tests have been conducted in the Grayburg Limestone reservoir in the Permian Basin, Texas,[37] and Austin Chalk and Buda Limestone Formations of the Darst Creek and Salt Flat Fields, Texas.[39] In the case of the Austin and Buda Formations, only the first and second cycles were successful. The causes of the failure of subsequent cycles were not determined at the time of the report; however, it was concluded that the failure was most likely owing to (a) insufficient energy to expel the released oil to the wellbore, (b) uncontrolled water entry into zones of high water saturation, and (c) wellbore damage. It was also concluded that the process would have been applicable to the Buda and Austin Formations if the above problems did not exist. The cost of obtaining the necessary oil saturation data and maintaining adequate control over injection, however, may be high enough to prohibit the use of pressure-pulsing process in complex marginal reservoirs.

Vugular–Solution Porosity Systems

The carbonate reservoirs with a vugular–solution porosity system exhibit a wide range of permeabilities. The permeability distribution may be relatively uniform or quite irregular. The reservoirs with a uniform permeability distribution will probably respond to waterflooding in a manner similar to that of reservoirs with an intercrystalline–intergranular porosity system. The reservoirs with the irregular permeability distributions may respond in a manner similar to that of the reservoirs having a fracture–matrix porosity system.

The Pegasus Ellenburger Field, Texas, is a good example of a vugular reservoir which responds to waterflooding in a similar manner as the reservoirs having a fracture–matrix porosity system.[40] The vugular porosity system of the Ellenburger is, however, due to the random occurrence of breccia fragments rather than to solution processes. The peripheral injection system which was used initially could not supply sufficient energy, and, consequently, the reservoir pressure in the center of the field continued to decline. To overcome this problem, two injectors were placed in the center. As a result, the pressure decline was arrested, but severe channeling was experienced and poor injectivity profiles were obtained. Thus, the water injection program was severely curtailed.[40]

The Canyon Reef reservoir, Texas, exhibits a more uniform distribution of permeabilities.[41] In 1955, a peripheral water injection system was installed in the Sharon Ridge Canyon Unit, which is one of several fields producing from the Canyon Reef. Injectors were deepened prior to the start of the flood to permit injection into the aquifer as well as laterally into the oil zone. On the average, 48% of the injected water entered the formation above and 52% below the oil–water contact. About 100% of the reef's gross thickness was open to the injection. In 1964, 9 years after the start of injection, the flood was considered successful. Recoveries were running slightly better than indicated by the Stiles prediction calculations. Black and Lacik[41] estimated that the ultimate waterflood recovery would be near 50% or twice the primary ultimate recovery of 25.2%.

Arps[42] presented the "permeability-block" method of waterflood prediction, which is a simplified version of the Stiles technique. Table 4-4 shows the details of the application of this method to the Tensleep Sandstone reservoir in Wyoming. It seems that this procedure could be utilized for the carbonate reservoirs having vugular–solution porosity system only when the permeability distribution is relatively uniform. A larger number of "blocks" may be required, however; this will provide greater detail in the WOR versus recovery curve. The Stiles assumptions listed on p. 87 hold true for the permeability-block method. An explanation of some of the calculations shown in Table 4-4 follows:

(10) Unit recovery factor is calculated with the following equation:

$$\text{WR} = 7758 \, \phi \left(\frac{1 - S_{wi}}{B_o} - S_{or} \right) \qquad (4\text{-}10)$$

TABLE 4-4

Computation of the Water-Drive Recovery Factor for Tensleep Sandstone Reservoir in Wyoming by the Permeability-Block or Modified Stiles Method[a]

		Group					Total
		1	2	3	4	5	
(1)	Permeability range, md	>100	50–100	25–50	10–25	0–10	
(2)	Fraction of samples	.085	.109	.145	.212	.449	
(3)	Average permeability, md	181.3	69.0	34.4	16.1	2.4	
(4)	Capacity in darcy-feet	1.543	.752	.499	.341	.108	
(5)	Average porosity (ϕ), fraction	.159	.150	.152	.130	.099	
(6)	Average resid. oil saturation (S_o), fraction	.173	.195	.200	.217	.222	
(7)	Relative water perm. (k_{rw})	.65	.63	.60	.56	.54	
(8)	Average interstitial water saturation (S_w), fraction	.185	.154	.131	.107	.185	
(9)	Relative oil perm. (k_{ro})	.475	.53	.61	.66	.47	
(10)	Est. unit recovery factor, bbl/acre-ft	725	693	721	623	414	
(11)	Cumulative wet cap. $= \Sigma$ (4)	1.543	2.295	2.794	3.135	3.243	
(12)	Cumulative clean oil cap. $= 3.243 - (11)$	1.700	0.948	0.449	0.108	0	
(13)	Water–oil ratio	15.5	36.0	76.5	307.9	∞	
(14)	Cum. rec.,[b] WOR = 15.5 Min. $k_{wet} = 100$ md	61.6	52.1	35.9	21.3	4.5	175.4
(15)	Cum. rec.,[b] WOR = 36.0 Min. $k_{wet} = 50$ md	61.6	75.5	71.9	42.5	8.9	260.4
(16)	Cum. rec.,[b] WOR = 76.5 Min. $k_{wet} = 25$ md	61.6	75.5	104.5	85.1	17.8	344.5
(17)	Cum. rec.,[b] WOR = 307.9 Min. $k_{wet} = 10$ md	61.6	75.5	104.5	132.1	44.6	418.3
(18)	Cum. rec.,[b] WOR = ∞ Min. $k_{wet} = 0$	61.6	75.5	104.5	132.1	185.9	559.6

[a] After Arps,[42] courtesy of the SPE of AIME.
[b] Cumulative recovery in bbl/acre-ft.

where WR = unit recovery factor, bbl/acre-ft; ϕ = porosity, fraction; S_{wi} = interstitial water, fraction; S_{or} = residual oil saturation, fraction; B_o = oil formation volume factor, bbl/STB.

(13) The water–oil ratio of the produced stream is calculated using the following equation:

$$\text{WOR} = \frac{\mu_o}{\mu_w} \frac{k_{rw}}{k_{ro}} \frac{\Sigma C_w}{\Sigma C_o} \qquad (4\text{-}11)$$

where μ_o = oil viscosity, cp; μ_w = water viscosity, cp; k_{rw} = relative permeability to water at residual oil saturation, md; k_{ro} = relative permeability to oil at interstitial water saturation, md; ΣC_w = cumulative wet capacity (Line 11), d-ft; ΣC_o = cumulative oil capacity (Line 12), d-ft.

(14) Cumulative oil recovery in bbl/acre-ft when Group 1 is watered out. For Group 1 the recovery is the product of Line 2 and Line 10. For the other groups, the recovery is the product of Line 2 and Line 10, multiplied by the ratio of its average permeability (Line 3) to the minimum wet permeability.

(15)–(18) Cumulative oil recovery in bbl/acre-ft as Groups 2 to 5 water out. As each group waters out, its recovery becomes simply (Line 2) × (Line 10). Recoveries for all other groups are multiplied by the ratio of the average permeability to the prevailing minimum wet permeability for each group.

The recovery is then plotted versus WOR. The economic limit WOR is calculated and the corresponding recovery is read from the curve. Unless the continuity of the permeability blocks across the reservoir is definitely established, the calculated recovery should be considered conservative.[42]

In carbonate pools, producing under a bottom water drive (natural or artificial), such as some of the vugular D-3 Reef reservoirs in Alberta, extreme ranges of permeabilities are found. Here the permeability-block method yields results that are much too low. This is caused by two factors: buoyancy and imbibition.[42]

The water advances through the highly permeable sections of the reef and bypasses oil in the tighter sections. When the bypassing occurs, a buoyancy gradient is set up across the tight sections due to the density difference between the oil and water, and this gradient tends to drive the oil into the more permeable sections. The imbibition of water from the high permeability areas into the tighter sections adds to the buoyancy effect.

The SACROC Unit in Kelly Snyder Field, Texas, is the site of an unusual approach to waterflooding of a reef (vugular–solution porosity system).[43] The injectors are run across the crest of the reef. This is not as extreme a procedure as seems at first. The Canyon Reef reservoir has a low mound shape and gently dipping flanks. Its thickness varies from 0 to 795 ft, and it is 7 miles wide. The use of a peripheral injection system was not feasible because the zone is thin at the edges. Allen and Thomas[43] list several advantages for the center-to-edge line drive flood: (a) The water moves in the same direction as that of the natural pressure gradients, (b) the water is injected into the thickest portion of the reef, permitting injection of large volumes of water and reducing the number of injectors required, and (c) the injection system costs less than a peripheral system due to fewer injectors, shorter injection line, and fewer pump stations. After 5 years of operation (reported in 1959) the flood was considered a success. The ultimate recovery was estimated at over twice the primary recovery of 23.6%.

Questions and Problems

4-1. List at least two types of predictive techniques that are applicable for each type of porosity system. List the limitations of each method proposed.

4-2. (a) List the assumptions common to most multilayered methods. (b) List at least three predictive techniques that can be used for a multilayered system.

4-3. Utilizing the data for Reservoir A, Appendix H, calculate the anticipated recovery by use of the Craig–Geffen–Morse method.

4-4. Utilizing the data for Reservoir A, Appendix H, calculate the anticipated recovery by use of the Craig–Stiles method.

4-5. List problems involved in predicting secondary recovery of oil from reservoirs with a well-developed fracture–matrix porosity system.

4-6. What reservoir characteristics give rise to strong imbibition effects?

4-7. Briefly describe the following: (a) cyclic injection, (b) pressure-pulsing injection, (c) slow-continuous injection, and (d) the advantages of each.

4-8. Prepare a flow sheet showing the data that must be gathered in order to use the Buckley–Leverett prediction technique for a carbonate reservoir having an intercrystalline–intergranular porosity system. Also state where this information can be obtained.

References

1. Goolsby, J. L. and Anderson, R. C.: "Pilot Waterflooding in a Dolomite Reservoir, The McElroy Field", *J. Pet. Tech.* (Dec., 1964) 1345–1350.
2. Calhoun, J. C., Jr., McCarthy, J. C. and Morse, R. A.: "Effects of Permeability on Secondary Recovery of Oil", in: *Secondary Recovery of Oil in the United States*, 2nd Ed., API (1950) 214–221.
3. Henry, J. C. and Moring, J. D.: "Pilot Waterflood Evaluation—Pandhandle Field", SPE preprint 1801 presented at the SPE Regional Secondary Recovery Symposium in Pampa, Texas (Oct. 26–27, 1967).
4. Abernathy, B. F.: "Waterflood Prediction Methods Compared to Pilot Performance in Carbonate Reservoirs", *J. Pet. Tech.* (March, 1964) 276–282.
5. Stiles, W. E.: "Use of Permeability Distribution in Waterflood Calculations", *Trans.*, AIME (1949) **186**, 9–13.
6. Craig, F. F., Geffen, T. M. and Morse, R. A.: "Oil Recovery Performance of Pattern Gas or Water Injection Operations from Model Tests", *Trans.*, AIME (1955) **204**, 7–15.
7. Hendrickson, G. E.: "History of the Welch Field San Andres Pilot Waterflood", *J. Pet. Tech.* (Aug., 1961) 745–748.
8. Habermann, B.: "The Efficiency of Miscible Displacement as a Function of Mobility Ratio", *Trans.*, AIME (1960) **219**, 264–272.
9. Caudle, B. H. and Witte, M. D.: "Production Potential Changes During Sweep-Out in a Five-Spot System", *Trans.*, AIME (1959) **216**, 446–448.
10. Muskat, M.: *Physical Principles of Oil Production*, McGraw-Hill, New York (1950).
11. Park, R. A.: "Pressure Maintenance by Waterflooding North Virden Scallion Field, Manitoba", SPE paper 1321 presented at the SPE Regional Meeting in Bakersfield, California (Nov. 4, 1965).
12. Caudle, B. H., Erickson, R. A. and Slobod, R. L.: "The Encroachment of Injected Fluids Beyond Normal Well Pattern", *J. Pet. Tech.* (May, 1955) 79–85.
13. Dyes, A. B., Caudle, B. H. and Erickson, R. A.: "Oil Production after Breakthrough as Influenced by Mobility Ratio", *J. Pet. Tech.* (April, 1954) 27–32.
14. Bleakley, W. B.: "Computers Calculate Flood Potential", *Oil and Gas J.* (Oct. 6, 1969) 147–149.
15. Akins, D. W., Jr.: "Primary High Pressure Waterflooding in the Pettit Lime Haynesville Field", *Trans.*, AIME (1951) **192**, 239–248.
16. (Editorial): "Rangely Waterflood", *Oil and Gas J.* (March 2, 1964).
17. Bleakley, W. B.: "Sound Planning = Successful Flood", *Oil and Gas J.* (March 8, 1969) 154–159.
18. Wilson, J. F.: "Waterflooding—Down Structure Displacement in Presence of a Gas Cap", *J. Pet. Tech.* (Dec., 1962) 1383–1388.
19. Miller, F. H. and Perkins, A.: "Feasibility of Flooding Thin, Tight Limestones", *Pet. Eng.* (April, 1960) 55–75.
20. Gealy, F. D., Jr.: "North Foster Unit—Evaluation and Control of a Grayburg–San Andres Waterflood Based on a Primary Oil Production and Waterflood Response", SPE preprint 1474 presented at the SPE 41st Annual Fall Meeting in Dallas, Texas (Oct. 2–5, 1966).
21. O'Briant, J. F.: "Operation and Performance Review of the Goldsmith–Cummins (San Andres) Unit Water Flood", *Proceedings of the Southwest Petroleum Short Course* (1967) 43–51.

22. Allen, W. W., Herriot, H. P. and Stiehler, R. D.: "History and Performance Prediction of Umm Farud Field, Libya", *J. Pet. Tech.* (May, 1969) 570–578.
23. Wood, B. O. and McShane, J. B., Jr.: "A Successful Glorieta–San Angelo Waterflood, Snyder Field, Howard County, Texas", *Proceedings of the Southwest Petroleum Short Course* (1969) 49–57.
24. Trube, A. S., Jr.: "Oil Production by Primary Artificial Frontal Water Drives in the New Hope Field, Franklin County, Texas", *Proceedings of Seventh Oil Recovery Conference*, Texas Petroleum Research Committee (May 7, 1954) 57–75.
25. Trube, A. S., Jr. and DeWitt, S. N.: "High Pressure Water Injection for Maintaining Reservoir Pressures, New Hope Field, Franklin County, Texas", *Trans.*, AIME (1950) **189,** 325–334.
26. Fickert, W. E.: "Economics of Water Flooding the Grayburg Dolomite in South Cowden Field", *Proceedings of the Southwest Petroleum Short Course* (1965) 21–31.
27. Wayhan, D. A. and McCaleb, J. A.: "Elk Basin Madison Heterogeneity—Its Influence on Performance", *J. Pet. Tech.* (Feb., 1969) 153–159.
28. Borgan, R. L., Frank, J. R. and Talkington, G. E.: "Pressure Maintenance by Bottom-Water Injection in a Massive San Andres Dolomite Reservoir," *J. Pet. Tech.* (Aug., 1965) 883–888.
29. Elkins, L. F.: "Internal Anatomy of a Tight, Fractured Hunton Lime Reservoir Revealed by Performance—West Edmond Field", *J. Pet. Tech.* (Feb., 1969) 221–232.
30. Willingham, R. W. and McCaleb, J. A.: "The Influence of Geologic Heterogeneities on Secondary Recovery from the Permian Phosphoria Reservoir, Cotton Creek, Wyoming", SPE preprint 1770 presented at the Rocky Mountain Regional SPE Meeting in Casper, Wyoming (May 22–23, 1967).
31. Graham, J. W. and Richardson, J. G.: "Theory and Application of Imbibition Phenomena in Recovery of Oil", *J. Pet. Tech.* (Feb., 1959) 65–69.
32. Brownscombe, E. R. and Dyes, A. B.: "Water-Imbibition Displacement—A Possibility for the Spraberry", *Drill. and Prod. Pract.*, API (1952) 383–390.
33. Barfield, E. C., Jordan, J. K. and Moore, W. D.: "An Analysis of Large Scale Flooding in Fractured Spraberry Trend Area Reservoir", *J. Pet. Tech.* (April, 1959) 15–19.
34. Guidroz, G. M.: "E. T. O'Daniel Project—A Successful Spraberry Flood", *J. Pet. Tech.* (Sept., 1967) 1137–1140.
35. Elkins, L. F. and Skov, A. M.: "Cyclic Waterflooding the Spraberry Utilizes 'End Effects' to Increase Oil Production Rate", *J. Pet. Tech.* (April, 1963) 877–884.
36. Elkins, L. F., Skov, A. M. and Gould, R. C.: "Progress Report on Spraberry Waterflood Reservoir Performance, Well Stimulation and Water Treating and Handling", *J. Pet. Tech.* (Sept., 1968) 1039–1049.
37. Owens, W. W. and Archer, D. L.: "Waterflood Pressure-Pulsing for Fractured Reservoirs", *J. Pet. Tech.* (June, 1966) 745–752.
38. Felsenthal, M. and Ferrell, H. H.: "Pressure Pulsing—An Improved Method of Waterflooding Fractured Reservoirs", SPE paper 1788 presented at SPE Permian Basin Oil Recovery Conference, Midland, Texas (May 8, 1967).
39. Hester, C. T., Walker, J. W. and Sawyer, G. H.: "Oil Recovery by Imbibition Water Flooding in the Austin and Buda Formations", *J. Pet. Tech.* (Aug., 1965) 919–925.
40. Cargile, L. L.: "A Case History of the Pegasus Ellenburger Reservoir", *J. Pet. Tech.* (Oct., 1969) 1330–1336.

41. Black, J. L. and Lacik, H. A.: "History of a Scurry County, Texas, Reef Unit", *Proceedings of the Southwest Petroleum Short Course* (1964) 35–39.
42. Arps, J. J.: "Estimation of Primary Oil Reserves", *Trans.*, AIME (1956) **207**, 182–191.
43. Allen, H. H. and Thomas, J. B.: "Pressure Maintenance in SACROC Unit Operations, January 1, 1959", *J. Pet. Tech.* (Nov., 1959) 42–48.

CHAPTER 5

Gas Injection—Immiscible Displacement

Introduction

Gas injection may be either a miscible or an immiscible displacement process. The character of the oil and gas and the temperature–pressure conditions of the injection determine the type of process. Only the immiscible displacement process is discussed in this chapter. Gas can be injected into a reservoir to maintain the pressure (i.e., dispersed gas injection) or to attempt to bank and sweep oil to the producers (i.e., gas cap injection). Both systems are considered in this chapter.

Theoretical calculations by Muskat[1] showed that oil recovery could be greatly increased by pressure maintenance particularly in the case of oils having a high formation volume factor. In these calculations, however, many physical characteristics of the heterogeneous reservoir were not taken into consideration. Elkins and Cooke[2] noted that the volume of gas injected and the associated change in oil viscosity and formation volume factor appear to be the principal factors in determining the oil recovery. To obtain the maximum benefit from the change in oil properties, the volume of reservoir contacted by the gas must be as high as possible. For massive carbonate reservoirs with good permeability, injection pressure does not appear to have a large effect on final ultimate oil recovery.[2]

The primary problems with gas injection in carbonate reservoirs are the high mobility of the displacing fluid and the wide variations in permeability. Much greater control over the injection process must be exercised than is necessary with waterflooding. The effect of reservoir's permeability profile on gas sweep efficiency may be evaluated most satisfactorily by a short-term pilot gas-injection test. The pilot test may also provide necessary data to calculate the required volumes of gas; this, in turn, will aid in the design of compressor equipment and estimating the number of injection wells which will be required. The following four recommendations should be followed in conducting a gas pilot test.[2]

1. The rate of gas injection and its duration should be sufficient to result in significant changes in produced gas–oil ratio. The daily injec-

tion rate should at least equal the volume of fluids produced by the first-row producing wells.
2. Gauging facilities should be adequate to ensure accurate measurement of the performance of the first-row wells. Continuous monitoring is preferred.
3. The test period should include a time period prior to and after the injection of gas to establish the trend of the produced oil–gas ratio.
4. When possible, the pilot test should be conducted early in the life of the field.

In the present chapter, useful design techniques are presented, followed by summaries of field applications. Analogy must be used with considerable care owing to the heterogeneity of the carbonate reservoir rocks and the sensitivity of the process.

Predictive Techniques

Two major predictive techniques, the Welge[3] and the Tarner,[4] are reviewed in this chapter. Either technique is applicable to the analysis of carbonate reservoirs.

Welge Method

The Welge technique[3] is a simplified form of the Buckley–Leverett or frontal advance method. It can also be used in evaluating waterflooding operations. The advantage of Welge's method over the Buckley–Leverett technique is that the curve showing the saturation distribution in the reservoir is not required in the former method. The average gas saturation is used to determine the recovery, whereas the terminal gas saturation near the outflow face defines the fractional flow of oil, f_o, and hence the flowing gas–oil ratio at any desired time during production. As a result, a considerable amount of time required for calculation and plotting is saved.

Assumptions
1. The reservoir is a single homogeneous layer, and the flow is linear.
2. Gas displaces oil saturated with gas at a constant pressure.
3. Oil displacement is immiscible.
4. Flow occurs in one direction only (there is no crossflow).

Procedure
1. Review the relative permeability data obtained from core analysis. If more than one set of data is available, choose a set which appears to

TABLE 5-1

Recommended Steps in Using the Welge[3] Technique of Gas-Injection Flood Prediction

(1)	(2)	(3)	(4)	(5)	(6)	(7)	(8)	(9)
S_g	k_g/k_o	k_g	f_o	df/dS	t	S_{ga}	GOR	q_o

be most representative. If stratification exists in the reservoir, a separate calculation for each layer should be considered. (See the discussion of the layered Buckley–Leverett technique, p. 77.)

2. Approximate the irregular shape of the reservoir with a linear model. An idealized length and cross-sectional area of the reservoir to flow must be determined.
3. Obtain the average fluid velocity, v_o, in the hydrocarbon-occupied pore space as follows: (a) An average cross-sectional area exposed to flow, A_f, is obtained by dividing the original oil-in-place (converted from bbl to cu ft) by the idealized length, L (ft), of the reservoir. (b) The v_o is then equal to the average withdrawal rate of oil, q_o, and water, q_w, divided by A_f:

$$v_o = \frac{q_o + q_w}{A_f} \tag{5-1}$$

The units of v_o should be converted from ft/day to cm/sec.

4. The remainder of the steps in the calculations can be presented in tabular form as shown in Table 5-1. The values for the columns in the table can be determined as described here.
    a. Column 1: several selected values of producing gas saturations, S_g, after breakthrough.
    b. Column 2: value of k_g/k_o for the selected S_g as determined from a k_g/k_o versus S_g curve.
    c. Column 3: values of k_g for each value of S_g selected.
    d. Column 4: flow of oil

$$f_o = \frac{1 + 3.09 \times 10^6\{[kk_{ro}(\rho_o - \rho_g)(\sin\theta)]/\mu_g v_o\}}{1 + (k_{rg}/k_{ro})(\mu_o/\mu_g)} \tag{5-2}$$

where f_o = fraction of oil in flowing stream; k_{rg} = relative permeability to gas with connate water in place; k_{ro} = relative permeability to oil with connate water in place; k = total permeability to oil with connate water in place, md; μ_g = gas viscosity, cp; μ_o = oil viscosity, cp; ρ_o = oil density, g/cm³; ρ_g = gas density, g/cm³; θ = angle of structural dip; v_o = average fluid velocity, cm/sec. Note that where dip is negligible, $\sin\theta$ approaches zero and Eq. 5-2 simplifies to

$$f_o = \frac{1}{1 + (k_{rg}/k_{ro})(\mu_o/\mu_g)} \tag{5-3}$$

e. Column 5: the values of df/dS are taken from f_o versus S_g curves similar to the f_w versus S_w curve presented in Fig. 3-22. The slopes of the curve, df/dS, may be measured directly for the desired saturations.

f. Column 6: the time, t, may be determined as follows:

$$t = \frac{L}{v_o(df/dS)} \tag{5-4}$$

where L = length of reservoir, cm (1 ft = 30.48 cm).

g. Column 7: average gas saturation, S_{ga}, of the reservoir is

$$S_{ga} = \left(S_g + \frac{f_o}{(df/dS)}\right) \tag{5-5}$$

h. Column 8: the producing gas–oil ratio, GOR, for various values of S_g is equal to

$$\text{GOR} = R_s + \left(\frac{B_o}{B_g}\right)\left(\frac{1-f_o}{f_o}\right) \tag{5-6}$$

where R_s = gas in solution per bbl of oil, cu ft/bbl; $(B_o/B_g) \times [(1-f_o)/f_o]$ = free gas produced per bbl of oil, cu ft/bbl; B_o = oil formation volume factor, bbl/STB; B_g = gas formation volume factor.

i. Column 9: the stock tank oil producing rate, q_o,

$$q_o = \frac{(v_o)(A_f)(\phi)(1-S_w)(f_o)}{5.615\,B_o} \tag{5-7}$$

The distance, X, from the injector that the flood front has traveled can be found at any time, t, on using the following equation:

$$X = (v_o)(t)\left(\frac{df}{dS}\right) \tag{5-8}$$

Equation 5-6 is sensitive to the assumed conditions such as gas velocity, permeability, and geometric configuration of the reservoir. It tends to yield GOR values higher than those occurring in the field for the same cumulative recovery. Consequently, if a GOR cutoff is based on this calculation, a conservative reserve estimate is obtained.

This method is sensitive to the throughput velocity, v_o. Inasmuch as the velocity is in the denominator of the gravitational term (Col. 4), rapid displacement reduces the size of the gravitational term and increases the fraction of gas, f_g, flowing through the reservoir. This would indicate a lower displacement efficiency. On the other hand, where the gravitational term is sufficiently large, f_g can approach zero or have negative values; this implies countercurrent flow of gas up dip and oil down dip resulting in gravity drainage which yields the maximum displacement efficiency. If gas cap overlies the oil zone, drainage can be vertical (sin θ = 1.00) and the cross-sectional area to flow can be very large. Gravity drainage will yield high recoveries unless the vertical permeability is low.

Frequently a field has been produced for a period of time prior to initiation of gas injection. In this case, the calculation steps in the Welge predictive technique are similar to those previously presented, except that calendar time is used. The following procedure, presented in outline form, can be followed:

1. Determine the residual oil saturation, $S_{o\,1}$, at the time of flood:

$$S_{o\,1} = \left(\frac{B_{o\,1}}{B_{oi}}\right)\left(S_{oi} - \frac{N_p B_{oi}}{7758\,Ah\phi(1 - S_w)}\right) \qquad (5\text{-}9)$$

where:

$B_{o\,1}$ = oil formation volume factor at start of injection, bbl/STB.
B_{oi} = initial formation volume factor, bbl/STB.
N_p = oil produced in stock tank barrels.
Ah = reservoir volume, acre-ft.
ϕ = porosity, fraction.
S_w = interstitial water, fraction.
S_{oi} = initial oil saturation, fraction.

2. Determine the residual gas saturation, $S_{g\,1}$, at start of injection:

$$S_{g\,1} = (S_{oi} - S_{o\,1}) \qquad (5\text{-}10)$$

3. The time at which this saturation, $S_{g\,1}$, is reached, which is computed following the tabulation shown in Table 5-1, is then identified with the calendar date of the start of injection.

Gas Injection—Immiscible Displacement

4. The average gas saturation at later times is calculated as follows:

$$S_{ga} = \left[S_g + \frac{f_o}{df/dS}\left(\frac{B_{oi}}{B_{o\,1}}\right) \right] \qquad (5\text{-}11)$$

Tarner Prediction[4]

Tarner[4] has developed a mathematical approach for predicting reservoir performance with and without gas injection. His method employs the simultaneous solution of two gas–oil ratio equations. The first equation is based on material balance concepts, whereas the second equation is based on the relative permeability relationships.

Assumptions
1. The reservoir can be treated as a single unit.
2. There are no pressure gradients across the field; the pressure drops uniformly.
3. All gas must flow through the oil in the reservoir; no bypassing occurs.
4. The size of the oil zone does not change. There is no water influx or gravity drainage.
5. Representative k_g/k_o versus S_g data are available.
6. S_w is constant.
7. Prediction starts when the bubble point is reached. No free gas saturation exists in the reservoir.

Data Required
1. Oil- and gas-in-place initially.
2. Cumulative oil and gas production until the time when the bubble point pressure is reached, and production to the start of the prediction.
3. Curves relating the instantaneous produced gas–oil and average reservoir pressure to cumulative production (both curves must extend to the start of prediction period).
4. Curves relating reservoir pressure to (a) formation volume factors (oil, gas, total), (b) solution gas–oil ratio, and (c) viscosity (oil and gas).
5. Curve relating permeability ratio and total liquid saturation.

Permeability ratio data can be obtained from laboratory work or determined from the reservoir performance data. "Apparent" relationships developed from reservoir performance data, particularly fluctuations caused by changes in the producing rate, enable engineers to predict the performance to be expected with and without gas injection.[5]

Equations

The instantaneous produced gas–oil ratio from the material balance equation is given by the following equation:

$$R_2 = \left[\frac{(B_{t\,2} - B_{tb}) - \left(\frac{N_{p\,2}}{N}\right)(B_{t\,2}) + \left(\frac{N_{p\,2}}{N}\right)(R_{sb})(B_{g\,2})}{(B_{g\,2})\left[\left(\frac{N_{p\,2}}{N}\right) - \left(\frac{N_{p\,1}}{N}\right)\right]} \right.$$

$$\left. - \frac{\left(\frac{N_{p\,1}}{N}\right)(R_{p\,1})}{\left[\left(\frac{N_{p\,2}}{N}\right) - \left(\frac{N_{p\,1}}{N}\right)\right]} - \frac{R_1}{2} \right] \quad (5\text{-}12)$$

where:

R_2 = instantaneous producing gas–oil ratio at estimated pressure, SCF/STB.
R_1 = instantaneous producing gas–oil ratio at previous pressure, SCF/STB.
$R_{p\,1}$ = cumulative gas–oil ratio at previous pressure, SCF/STB.
R_{sb} = solution gas–oil ratio at bubble point, SCF/STB.
B_{tb} = total formation volume factor (two phase) at bubble point pressure, bbl/STB.
$B_{t\,2}$ = total formation volume factor (two phase) at estimated pressure, bbl/STB.
$B_{g\,2}$ = gas formation volume factor at estimated pressure, bbl/SCF.
N = oil-in-place at bubble point pressure, STB.
$N_{p\,2}$ = cumulative oil produced at estimated pressure, STB.
$N_{p\,1}$ = cumulative oil produced at previous pressure, STB.

The total liquid saturation at a pressure below the bubble point pressure is given by

$$S_{t\,2} = \left(1 - \frac{N_{p\,2}}{N}\right)\left(\frac{B_{o\,2}}{B_{tb}}\right)(1 - S_{wi}) + S_{wi} \quad (5\text{-}13)$$

where:

$S_{t\,2}$ = total liquid saturation at estimated pressure, percent.
S_{wi} = initial water saturation, percent.
$B_{o\,2}$ = oil formation volume factor at estimated pressure, bbl/STB.

Gas Injection—Immiscible Displacement

The instantaneous produced gas–oil ratio derived from Darcy's equation is given by

$$R_2 = \left(\frac{k_g}{k_o}\right)_2 \left(\frac{\mu_o}{\mu_g}\right)_2 \left(\frac{B_o}{B_g}\right)_2 + R_{s\,2} \qquad (5\text{-}14)$$

where:

$(k_g/k_o)_2$ = permeability ratio at estimated pressure.
$\mu_{o\,2}$ = oil viscosity at estimated pressure, cp.
$\mu_{g\,2}$ = gas viscosity at estimated pressure, cp.
$B_{g\,2}$ = gas formation volume factor at estimated pressure, bbl/SCF.
$R_{s\,2}$ = solution gas–oil ratio at estimated pressure, SCF/bbl.

The cumulative produced gas is given by

$$G_{p\,2} = \left[\frac{(R_1 + R_2)}{2}\right](N_{p\,2} - N_{p\,1})(1 - G_i) + (G_{p\,1}) \qquad (5\text{-}15)$$

where:

$G_{p\,2}$ = cumulative produced gas at estimated pressure, MCF.
$G_{p\,1}$ = cumulative produced gas at previous pressure, MCF.
G_i = injected gas (fraction of gas produced which is returned to the reservoir during the incremental period).

The cumulative produced gas–oil ratio, $R_{p\,2}$, in SCF/STB may be expressed as

$$R_{p\,2} = \frac{G_{p\,2}}{N_{p\,2}} \qquad (5\text{-}16)$$

Procedure

The two equations for the instantaneous produced gas–oil ratio are solved by use of the following method:

1. Calculate oil- and gas-in-place at bubble point pressure.
2. Assume an average reservoir pressure. Select pressure steps on the basis of the change in character of the instantaneous produced gas–oil ratio. (Caution: use smaller pressure increments where the slope of the curve is changing rapidly.)
3. Make three estimates of cumulative oil production for the assumed pressure (note the trend of the cumulative oil production–pressure plot when selecting values).
4. Calculate the instantaneous produced gas–oil ratios using Eqs. 5-12 through 5-14 for each of the three estimates of cumulative oil production.

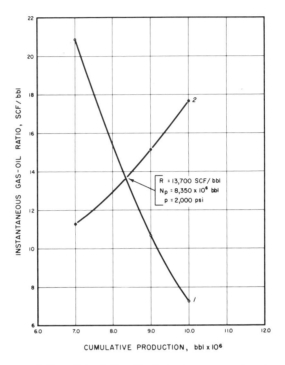

FIG. 5-1. Tarner prediction plot relating the instantaneous produced gas–oil ratio and the cumulative oil production (curve 1, performance calculated using the material balance equation; and curve 2, performance calculated from relative permeability data).

5. Plot the data as shown in Fig. 5-1. The intersection of the two curves gives the answer.
6. Calculate the cumulative produced gas using Eq. 5-15.
7. Calculate the cumulative produced gas–oil ratio using Eq. 5-16.

Reservoir Performance

Gas cycling may be distinguished from gas injection in that the recovery of the crude oil by vaporization is the primary intent of cycling operations. The intent of gas injection is to provide a pressure gradient which moves the oil toward nearby producers.

Gas Cycling

Gas cycling has been used successfully in several carbonate reservoirs: Pickton in Texas,[6] Harmattan-Elkton in Alberta, Canada,[7] and Opelika in Texas.[8] Cook et al.[9] demonstrated in the laboratory that oil recoveries after cycling operations could range from 15 to 70% of the immobile stock-tank oil depending upon the characteristics of the reservoir fluids. The total (primary plus secondary) recovery from the Pickton reservoir was estimated at near 73% (primary ultimate was estimated at 19.4%) of the oil originally in place.[6] The total recovery in the Harmattan-Elkton field was estimated at 30–40% of the oil-in-place.[7] The total recovery from the Opelika field was estimated at 27%.[8]

Table 5-2 and Figs. 5-2 and 5-3 show the theoretical effects of oil gravity, pressure, and temperature and the quantity of gas cycled upon the ultimate volume of oil recovered.[9] The viscosities for the oil and gas were obtained from data presented by Katz et al.[10] These data, however, cannot be extrapolated to field performance; they are presented only for the purpose of indicating the impact of various factors.

Jacoby and Berry[11] showed that the actual vaporization sequence occurring in each reservoir is determined by the production process. They outlined a method for calculating fluid compositions in reservoirs containing low-molecular-weight hydrocarbon liquids. The recovery of stock-tank oil per unit of pressure decline should be predicted from reservoir fluid composition (compositional material balance and oil–gas separator

TABLE 5-2

Effect of Oil Gravity, Pressure, and Temperature on Vaporization[a]

Example	Oil gravity, °API	Cycling pressure, psia	Cycling temperature, °F	Gas–oil viscosity ratio	Vaporization (percent of immobile stock-tank oil[b])	Displacement (percent of original stock-tank oil)
1	45	4100	250	0.338	73.6	76.5
2	35	4100	250	0.113	50.3	62.5
3	22	4100	250	0.020	28.9	54.0
4	35	4100	100	0.047	33.8	60.7
5	35	1100	100	0.007	15.3	48.8
6	45	1100	250	0.050	47.9	65.5

[a] After Cook et al.,[9] courtesy of the SPE of AIME.
[b] Amount of total gas cycled was 25 MCF/bbl of immobile stock-tank oil.

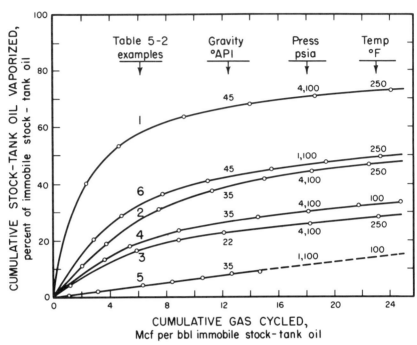

Fig. 5-2. Vaporization as related to gravity of oil, pressure, temperature, and amount of gas cycled. (After Cook et al.,[9] courtesy of the SPE of AIME.)

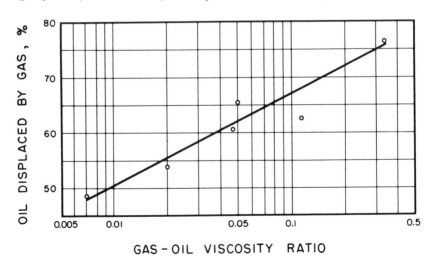

Fig. 5-3. Relationship between gas-oil viscosity ratio and oil displacement. (After Cook et al.,[9] courtesy of the SPE of AIME.)

recovery calculations). Jacoby and Berry[11] noted that for an oil rich in light components, the material balance method indicated a recovery twice that obtained on using conventional calculations.

Gas Injection

Injection of gas in carbonate reservoirs can increase the ultimate recovery of oil from the reservoir. The benefits obtained by the gas injection is dependent upon horizontal and vertical sweep efficiency of the injected gas. The sweep efficiency depends on the type of porosity system present, as discussed below.

Intercrystalline–Intergranular Porosity Systems

According to Longren,[12] insufficient data prohibited a quantitative prediction of the susceptibility to gas injection of the Panhandle Dolomite Reservoir, Texas, which is classified as an intercrystalline–intergranular porosity system. Secondary recovery predictions for the West Pampa Project in the Panhandle Field, Texas, evolved through pilot testing. Particular attention was paid to the selection and preparation of old producers for gas injection, the determination of maximum injection rates, and the evaluation of injection response.[12]

The primary concern, in deciding which producers should be used as injectors, was to obtain maximum displacement efficiency. Wells located in the areas having lower permeability were chosen in preference to those located in areas with higher permeability, because reservoir pressure was higher in the former areas and thus helped retard the fingering of gas. Preference was also given to the wells with low gas–oil ratios and those located in areas having lower gas saturations. Producers which penetrated the entire producing zone provided the best vertical coverage. Deepening of producers prior to the initiation of injection helped improve vertical coverage. The injectors located in areas having higher gas saturations required higher quantities of injected gas. Structural position did not appear to be a major factor affecting the performance of the injectors.

It was found that the maximum benefit from gas injection was derived when the producing wellbores were kept in clean condition and each producer was offset by at least one injection well.[13] Well cleanout and acidizing programs, performed after gas injection was underway for a year or more, yielded better results than when stimulation work was done in producers prior to gas repressurization.

Fracture–Matrix Porosity System

Representative data collection is difficult for reservoirs with a well-developed fracture–matrix porosity system consisting of low-permeability matrix and interconnected fractures and solution channels constituting 10% or more of the reservoir's void space.[14] As a result, calculations of recoverable oil by volumetric measurements will seldom yield volumes equal to those determined by material balance calculations. This divergence is greater where the permeability of the reservoir matrix is low.

Factors directly responsible for this difference are (1) significant differences between the pressure in the tight rock matrix and pressure in the interconnected fractures and/or solution channels; (2) difficulties in determining the quantity and distribution of the gas saturation within the rock matrix, fractures, and solution channels (it should be noted that not all of the matrix in a tight reservoir need be saturated with oil[14]); and (3) the quantity and rate of water influx during repressurization.

A well-developed fracture–matrix porosity system can result in severe channeling of injected gas and/or water influx with little or no benefit to the ultimate recovery of oil. An example of this occurred in the Bois d'Arc Formation pilot in the West Edmond Field, Texas.[14]

To aid in the determination of the reservoir volume swept by gas in the Hunton Lime, West Edmond Field, helium was added for one week to the injected gas as a tracer (1% concentration).[14] The producing wells were monitored for the daily quantity of produced helium for 5 months. At the end of 5 months, 44% of the injected helium tracer had been produced. The quantity of produced helium was affected by factors such as (1) varying velocities along various flow paths, (2) dispersion of gas along flow paths, and (3) variation in permeability. If the peak helium contents, however, correspond to average travel times of injected gas to producing wells, then the minimum percentage of reservoir volume swept by gas was about 6%. The additional oil recovered by gas injection in the Clearfork Zone dolomite, Fullerton Field, Texas, was around 4% of the total oil-in-place.[15] For both fields, a well-developed fracture–matrix porosity system reduced the effectiveness of injected gas. It was demonstrated, however, that gas injection into a fractured reservoir could result in recovery of additional oil.

Vugular–Solution Porosity System

As a gas conservation measure and in order to increase the ultimate oil recovery in a vugular–solution porosity type system, gas was injected into the Westerose D-3 Pool, Alberta, Canada.[16] Based on reservoir pressure behavior, it was determined that all pools on the reef trend are, to a varying degree, in pressure communication with each other. To predict the performance of the Westerose D-3 Pool and other pools on this reef trend, a two-dimensional computer model was developed to simulate the pressure–production performance of the pools. The model parameters were adjusted until there was close agreement between the reservoir model and the 20-year production history.

The future performance was then estimated at various rates of oil production with and without the return of the produced gas to the gas caps. It should be emphasized that the results are only qualitative and not quantitative. The trends determined are likely to be valid, but the quantitative values calculated should not be considered exact. The results of the simulation studies indicated that the Westerose D-3 Pool had lost, and would continue to lose, considerable oil to the aquifer unless gas injection was stopped. As a result of this investigation, the gas injection project was shut down shortly after the simulation study was completed.

The failure of gas injection in the Westerose D-3 Pool does not mean that it cannot be successful in other reef reservoirs. It does emphasize, however, the need to understand the geology of the reservoir completely when selecting a method of secondary oil recovery.

Questions and Problems

5-1. What factors govern miscibility in a gas-injection process?
5-2. What information may be gained from a gas-injection pilot? What considerations should be given to the location of gas-injection pilots? Explain reasons.
5-3. Using Tarner's method and the data of Reservoir A, Appendix H, calculate the gain in oil recovery from gas injection.
5-4. Using Welge's method and the data of Reservoir A, Appendix H, calculate the gain in oil recovery by gas injection.
5-5. Discuss the factors that influence the volume of gas required in a cycling project.
5-6. What is the effect of wettability on the oil recovery by gas drive versus that by water drive?

References

1. Muskat, M.: *Flow of Homogeneous Fluids*, J. W. Edwards, Inc., Ann Arbor, Michigan (1946).
2. Elkins, L. F. and Cooke, J. T.: "Pilot Gas Injection—Its Conduct and Criteria for Evaluation", *Trans.*, AIME (1949) **148,** 180–188.
3. Welge, H. J.: "Simplified Method for Computing Oil Recovery by Gas or Water Drive", *Trans.*, AIME (1952) **195,** 91–98.
4. Tarner, J.: "How Different Size Gas Caps and Pressure Maintenance Programs Affect the Amount of Recoverable Oil", *Oil Weekly* (June 12, 1944) 32–34.

5. Elkins, L. F.: "The Importance of Injected Gas as a Driving Medium in Limestone Reservoirs as Indicated by Recent Gas-Injection Experiments and Reservoir-Performance History", in: *Secondary Recovery of Oil in the United States*, 2nd Ed., API (1950) 370–382.
6. McGraw, J. H. and Lohec, R. E.: "The Pickton Field—A Review of a Successful Gas Injection Project", *J. Pet. Tech.* (April, 1964) 399–405.
7. Donahoe, C. W. and Bohannon, D. L.: "Harmattan Elkton Field—A Case for Engineered Conservation and Management", *J. Pet. Tech.* (Oct., 1965) 1171–1178.
8. Clay, T. W.: "Pressure Maintenance by Gas Injection ... In Opelika Field of Henderson County, Texas", in: *Secondary Recovery, Oil and Gas J.* Reprint Series, 55–58.
9. Cook, A. B., Johnson, F. S., Spencer, G. B. and Bayazeed, A. F.: "The Role of Vaporization in High Percentage Oil Recovery by Pressure Maintenance", *J. Pet. Tech.* (Feb., 1967) 245–250.
10. Katz, D. L., Cornell, D., Riki, K., Poettmann, F. H., Vary, J. A., Elenbaas, J. R. and Weinany, C. F.: *Handbook of Natural Gas Engineering*, McGraw-Hill, New York (1959) 175–179.
11. Jacoby, R. H. and Berry, V. J., Jr.: "A Method for Predicting Pressure Maintenance Performance for Reservoirs Producing Volatile Crude Oil", *Trans.*, AIME (1958) **213**, 59–64.
12. Longren, H. F.: "Increasing Oil Recovery in the Panhandle", *Proceedings of Texas Petroleum Research Committee*, 2nd Conference (1951) 143–145.
13. Nesluge, F. J.: "Gas Injection in Dolomite Reservoir West Pampa Repressuring Association Project", *Proceedings of Texas Petroleum Research Committee*, 2nd Conference (1951) 119–142.
14. Elkins, L. E.: "Internal Anatomy of a Tight, Fractured Hunton Lime Reservoir Revealed by Performance—West Edmond Field", *J. Pet. Tech.* (Feb., 1969) 221–232.
15. Nolan, W. E. and Locker, G. R.: "How Fractured Limestone Responds to Gas Injection", in: *Secondary Recovery, Oil and Gas J.* Reprint Series, 59–63.
16. Hnatiuk, J. and Martinelli, J. W.: "The Relationship of the Westerose D–3 Pool to Other Pools on the Common Aquifer", *J. Can. Pet. Tech.* (April–June, 1967) 43–49.

CHAPTER 6
Miscible Flooding

Introduction

The recovery of oil is dependent upon the volume of the reservoir contacted by the displacing fluid, the oil content at the start of the process, and the displacement efficiency of the injected fluid in the reservoir contacted.

Recognizing that 100% displacement efficiency requires the elimination of the interfacial forces between the displacing and displaced fluids, researchers studied various approaches to the achievement of miscible displacement. Among the approaches studied were (1) high-pressure gas injection, (2) enriched gas injection, (3) injection of a liquid petroleum gas, LPG, slug followed by either gas or water, and (4) injection of an alcohol slug followed by water.

A major problem encountered in the field applications of the miscible processes has been the poor volumetric sweep efficiency which has generally limited the total oil recovery to 55–65% of the original oil-in-place.[1] The more important factors controlling the volumetric sweep efficiency are (1) the reservoir size and shape, (2) the irregularity of the porosity and permeability within the reservoir, (3) the structural dip of the reservoir, (4) the mobility ratio of the displacing to the displaced fluid, and (5) the existing fluid saturations in the reservoir at the start of the flood.[1,2]

A second major problem is the state proration statutes. Currently in those states which have proration allowables, a waterflood is given the same allowable as a miscible drive in most areas, even though the miscible processes are more expensive to utilize.[1] In Alberta, Canada, where allowables are based on reserves, the high recovery associated with the miscible processes would allow higher production rates. These processes are, therefore, generally more attractive in Alberta than in the United States.

Miscibility

Miscibility exists when two fluids are able to mix in all proportions without any interface forming between them.[2] Miscibility is controlled by

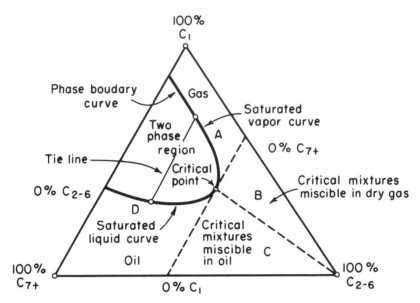

FIG. 6-1. Triangular graph showing physical conditions of hydrocarbon systems at fixed temperature and pressure conditions. (After Clark et al.,[2] courtesy of the SPE of AIME.)

the pressure and temperature, the composition of the oil, and the composition of the displacing fluid. The pseudo-ternary phase diagram is often used as an aid in understanding the miscibility process for complex hydrocarbon mixtures.[3] The diagram, however, should not be used for making quantitative predictions. Figure 6-1 shows a pseudo-ternary composition diagram from the work of Clark et al.[2] The three-component system shown consists of methane (C_1), the intermediates (C_2 through C_6) and all hydrocarbons heavier than C_6 (C_{7+}). The phase diagram shows whether gas, liquid, or gas–liquid phases exist for various mixtures of components at a given temperature and pressure. Region A represents an all-gas phase whereas region D is all liquid (oil). In the critical regions, B or C, both liquid and gas are present. Region B shows the range of compositions for a given temperature and pressure that would be miscible with the mixtures in the dry gas region. Region C contains mixtures that are miscible with the mixtures in the liquid (oil) region. The tie lines terminate at points on the saturated vapor curve and the saturated liquid curve. These two points represent a saturated gas and a saturated oil which are in equilibrium.[2] The sign of the slope for the tie line is determined by the value of the equilibrium ratios (constants) for the intermediates as the system

Miscible Flooding 147

approaches its critical composition.[3] If the equilibrium ratios total less than one, the slope is negative. If they total more than one, the slope is positive (the tie line in Fig. 6-1 has a positive slope). The limiting tie line passes through the critical point. All fluids having an intermediate composition equal to or greater than that of the critical composition are either immediately miscible or are capable of becoming miscible with the crude in region D.[3]

In a gas–liquid system, a miscible bank is formed by either evaporation or condensation of the intermediate hydrocarbons (C_2 through C_6). If the major transfer of intermediate hydrocarbons occurs by condensation from the gas, the system is known as enriched-gas or condensing-gas drive. If the major transfer of intermediate hydrocarbons is from the reservoir oil, then the system is known as high-pressure or evaporating-gas drive.

Sweep Efficiency

According to Crosby[4] many of the unsuccessful miscible projects were improperly engineered. Some of the reasons for failure were improper patterns, insufficient pressure to achieve miscible displacement, and excessive reservoir heterogeneity. When the conditions are suitable for the attainment of miscibility, sweep efficiency determines the success or failure of a project.

The areal coverage or pattern efficiency of a project is controlled by the shape of the reservoir and the location of the injection wells.[2] The location of the injection wells, the injection rates, and the timing of fluid production or injection should be evaluated by pilot operations.

Limestone reservoirs seldom are homogeneous. Lithology and petrophysical properties vary from one area to the other, and the reservoirs frequently tend to be stratified. As variations in the porosity and permeability increase, the sweep efficiency decreases.[2] These variations are less important in waterflooding owing to the benefit received by the water's ability to wet the sides of the pore channels and enter the channels of low permeability. In laboratory tests, Scott and Read[5] found that miscible displacement sweep efficiency was related directly to pore geometry, i.e., combined effects of tortuosity, pore shape, pore-size distribution, and width of intercommunicating channels.

Steeply dipping reservoirs or reefs with good vertical permeability can be effectively flooded when the impact of gravity is considered and the rates of injection are controlled.

In the case of reefs, good vertical drainage can result in oil recovery efficiency as high as 85–95% of the oil-in-place.[6] In massive and fractured

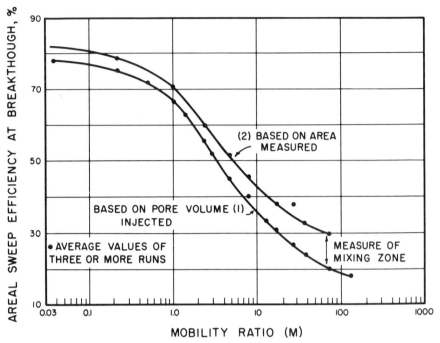

Fig. 6-2. Breakthrough sweep efficiencies for a five-spot well pattern as a function of mobility. The difference between areas contacted and the corresponding volumetric injection is a measure of the size of the mixing zone between the two miscible phases. (After Habermann,[8] courtesy of the SPE of AIME.)

reservoirs, the benefits of gravity segregation are generally eliminated by the presence of stratification.[7]

The impact of mobility ratio, M, on the efficiency of miscible displacement is shown in Figs. 6-2, 6-3, and 6-4.[8] When a less viscous fluid, such as a liquid petroleum gas (LPG) slug, drives a more viscous reservoir oil, fingers develop which reduce the areal sweep efficiency. For $M \leq 1$, the front is essentially radial, and a stable band is formed by the slug between the oil and driving phase until the drawdown of the producer causes a center finger to develop, which moves ahead to the well. For $M > 1$, viscous fingering of the miscible phase occurs sooner with the resultant decrease in oil recovery.[8]

High-Pressure Gas Injection

In the high-pressure gas injection process, a lean gas (low in the C_2 through C_6 hydrocarbons, i.e., intermediates) is utilized. The injected gas is enriched within the reservoir by a transfer of intermediate hydrocarbons from the oil to the gas. To accomplish this transfer, the oil must contain sufficient quantities of hydrocarbons in the C_2 through C_6 range, and the reservoir pressure should generally be in excess of 2500 psi. If the reservoir

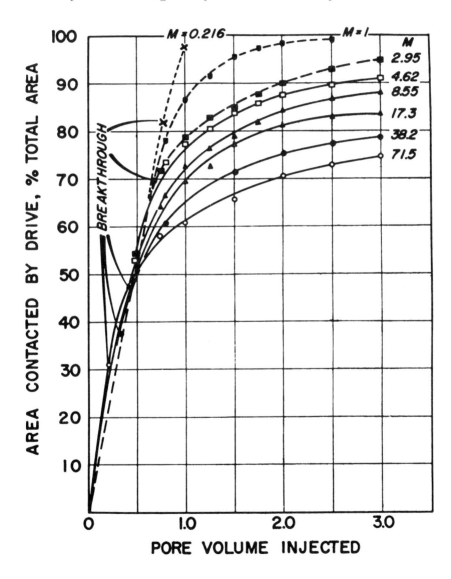

Fig. 6-3. Area contacted by drive after breakthrough, based on a quarter of a five-spot well pattern. (After Habermann,[8] courtesy of the SPE of AIME.)

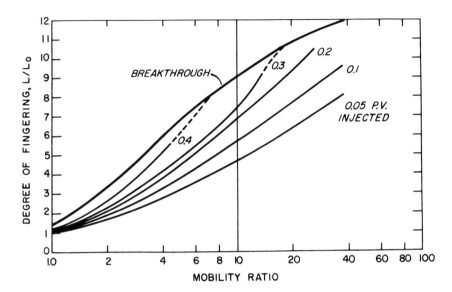

Fig. 6-4. Degree of fingering, based on a quarter of a five-spot well pattern. (After Habermann,[8] courtesy of the SPE of AIME.)

oil is low in intermediate hydrocarbons, it will be impossible for an evaporating-gas drive to develop a miscible front.[9]

The performance of a miscible flood may be approximated with the Buckley–Leverett or Welge predictive techniques when the following assumptions are added to those presented previously (p. 74): (1) Using the viscosity of the reservoir fluid and the viscosity of the displacing fluid, the recovery at breakthrough can be determined using the curves presented for liquid–liquid displacement (Fig. 4-4); (2) the producing gas–oil ratio can be estimated by assuming the fluids are produced in proportion to their saturation and viscosities; and (3) the time for reaction between the injected gas and the reservoir fluid can be obtained from laboratory testing (determining the ratio in which gas and the oil combine). The preceding assumptions are based on lecture notes of Prof. Paul Crawford (Texas A&M Advanced Reservoir Engineering short course, 1964).

Enriched-Gas Drive

An enriched-gas-drive system forms a miscible or solvent front by condensation of the intermediates from the injected enriched gas. Under

Miscible Flooding

reservoir conditions, the intermediates condense to form the solvent bank. The process is applicable in the general pressure range of 1700 to 3000 psi.

Benham et al.[3] prepared a set of curves relating the maximum methane content in the displacing fluid (to support miscibility) to reservoir and injection fluid properties, and to reservoir temperature and pressure (Figs. 6-5 through 6-15). In comparing different reservoir oils at different temperatures and pressures, the C_2^+ content of the reservoir fluid and the C_5^+ content of the displacing fluid were found to be the controlling factors for attainment of miscibility. Figures 6-5 to 6-15 are not applicable if the displacing gas contains large amounts of nonhydrocarbons. The attainment of miscibility using an enriched-gas-drive process is favored by higher

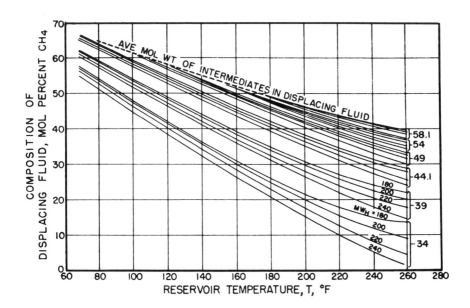

FIG. 6-5. Predicted phase conditions at 1500 psia for a miscible displacement of a reservoir fluid characterized by a C_{5+} molecular weight of 180, 200, 220, and 240. (After Benham et al.,[3] courtesy of the SPE of AIME.)

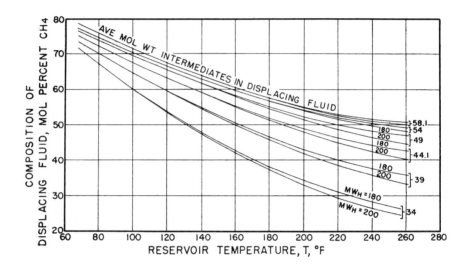

Fig. 6-6. Predicted phase conditions at 2000 psia for a miscible displacement of a reservoir fluid characterized by a C_{5+} molecular weight of 180 to 200. (After Benham et al.,[3] courtesy of the SPE of AIME.)

pressures, lower temperatures, light reservoir fluids, and displacing fluids having light hydrocarbons.[3]

The method of predicting maximum methane content to support miscibility for an enriched-gas drive can be summarized as follows[3]:

1. Calculate the molecular weight of the C_5^+ fraction of the reservoir fluid by multiplying the molecular weight of each component, C_5 or heavier, by its mole percent and summing.
2. Calculate the molecular weight of the C_2^+ fraction of displacing fluid in the same manner.
3. Determine the maximum methane content in the displacing fluid from the correlation charts for a particular reservoir pressure and temperature (Figs. 6-5 to 6-15).

Miscible Flooding

Volumetric prediction methods can be used in predicting oil recovery for enriched-gas-drive floods, if it is assumed that gravity effects may be neglected and that the injected dry gas does not move vertically.[10] Methods such as Stiles or Dykstra–Parsons permit calculation of a vertical sweep efficiency, E_v. Areal sweep efficiencies, E_a, may be estimated from Fig. 4-4, whereas the displacement efficiency, E_d, can be estimated as being 100% for the contacted area of the oil-saturated reservoir. The oil recovered, N_p, by miscible drive can be calculated as follows:

$$N_p = 7758Ah\phi(E_v E_a E_d) \frac{(1 - S_{wa} - S_{gi})(B_{oi} - B_{or})}{(B_{oi} B_{or})} \quad (6\text{-}1)$$

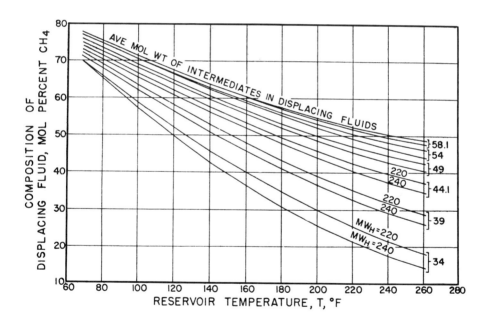

FIG. 6-7. Predicted phase conditions at 2000 psia for a miscible displacement of a reservoir fluid characterized by a C_{5+} molecular weight of 220 and 240. (After Benham et al.,[3] courtesy of the SPE of AIME.)

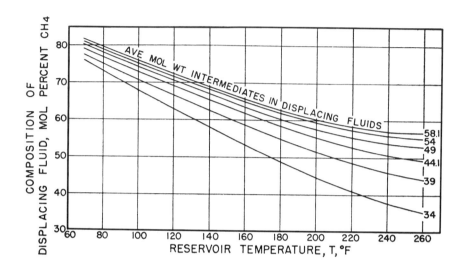

Fig. 6-8. Predicted phase conditions at 2500 psia for a miscible displacement of a reservoir fluid characterized by a C_{5+} molecular weight of 180. (After Benham et al.,[3] courtesy of the SPE of AIME.)

where:

A = area, acres.
B_{oi} = formation volume factor at beginning of project, bbl/STB.
B_{or} = formation volume factor at end of project, bbl/STB.
h = average reservoir thickness, ft.
ϕ = porosity, fraction.
S_{wa} = average formation water saturation, fraction.
S_{gi} = average gas saturation at start of project, fraction.

Welge et al.[11] have presented a method of predicting future recovery for a nonmiscible or partially miscible enriched-gas drive. Their method takes into account condensation of some of the intermediate hydrocarbons from the injected gas into the oil, as well as enhanced volatility of the heavier hydrocarbons at elevated temperatures. The recovery of oil is increased by the swelling of oil and the increase of the oil's mobility.

On the basis of laboratory experiments, Arnold et al.[12] concluded that in reservoirs, which are not subject to gross bypassing of the oil by the injected enriched gas, small banks of enriched gas driven by methane may be used to obtain a recovery comparable to that produced by continuous injection of enriched gas. The minimum bank size was estimated at 12% of the hydrocarbon pore volume.

Liquid Petroleum Gas (LPG) Slug Drive

In miscible slug injection, a slug or bank of LPG or propane is driven by dry gas or water through the reservoir. This slug miscibly displaces

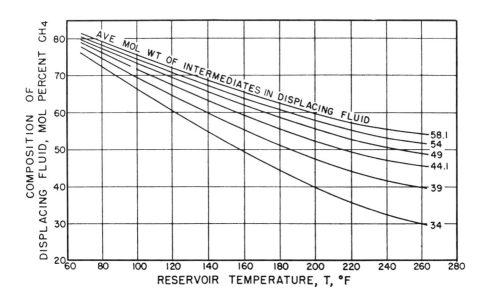

FIG. 6-9. Predicted phase conditions at 2500 psia for a miscible displacement of a reservoir fluid characterized by a C_{5+} molecular weight of 200. (After Benham et al.,[3] courtesy of the SPE of AIME.)

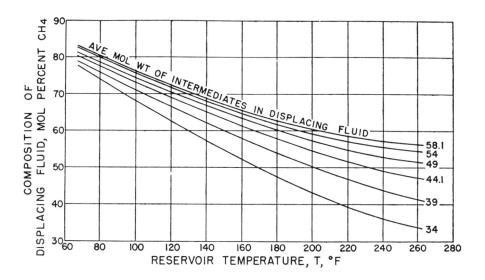

Fig. 6-10. Predicted phase conditions at 2500 psia for a miscible displacement of a reservoir fluid characterized by a C_{5+} molecular weight of 220. (After Benham et al.,[3] courtesy of the SPE of AIME.)

the reservoir oil from the swept portions of the reservoir. At pressures above 1100 psi, the LPG is also miscible with the driving gas.[13]

The quantity of LPG required to maintain miscible conditions is an important factor in the economics of miscible flooding. In the case of low solvent (LPG) content, miscibility is lost when the bank of LPG deteriorates. At that point, the displacement will become immiscible rather than miscible, and recovery will drop accordingly. On the other hand, an excess of LPG represents an unnecessary capital investment.

Factors such as (1) rock permeability, (2) displacement rate, (3) reservoir oil viscosity, (4) distance traveled, and (5) diffusion rate determine the extent of mixing which occurs at the solvent–crude oil interface and the solvent–driving gas interface.[13] Mixing tends to occur longitudinally in the direction of flow.

The specific quantity of slug required is a subject of controversy. Estimates range from 2–3% of the total pore volume[14] to 10–15% of the total pore volume.[8] Field trials indicate that the required slug size for many sandstone reservoirs is 10–15%, whereas carbonates appear to require 4–7% of the total pore volume. Habermann[8] investigated the effect of slug size (10–25% of the pore volume swept) upon the oil recovery. Using pore volume swept instead of total pore volume requires an estimation of vertical sweep efficiency, E_v, and the areal sweep efficiency, E_a, as previously discussed on p. 147. It was noted that for reservoirs with an intercrystalline–intergranular porosity system, when the slug size was less than 10% of the pore volume swept, the bank deteriorated prematurely, letting the driving phase penetrate through the bank of solvent. The

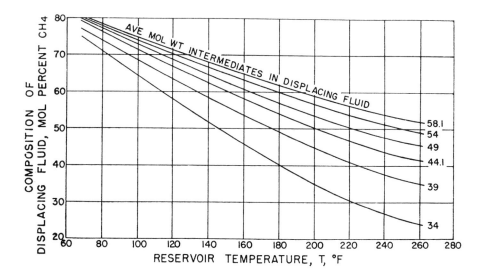

Fig. 6-11. Predicted phase conditions at 2500 psia for a miscible displacement of a reservoir fluid characterized by a C_{5+} molecular weight of 240. (After Benham et al.,[3] courtesy of the SPE of AIME.)

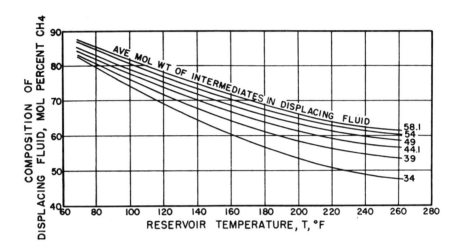

Fig. 6-12. Predicted phase conditions at 3000 psia for a miscible displacement of a reservoir fluid characterized by a C_{5+} molecular weight of 180. (After Benham et al.,[3] courtesy of the SPE of AIME.)

optimum size for the slug appeared to be about 15%.[8] Habermann's findings were substantiated by the work of Mahaffey et al.[7]

The sweep efficiency generally improves on using water as the driving fluid. There is, however, a loss of miscibility at the trailing edge of the slug, which leaves expensive solvent behind in the reservoir. To overcome this problem, some operators[15] inject a slug of propane, followed by a buffer zone of gas and then water.

Predictive Techniques

In addition to the techniques discussed previously, several computer predictive techniques have been reported in the petroleum industry literature. References 16 to 28 are presented for the reader who desires to investigate this subject in greater detail.

Intergranular–Intercrystalline Porosity System

A high-pressure gas miscible injection project was initiated in the Block 31 Field, Texas, in 1949.[29] In 1969, it was estimated that 60% of the oil-in-place would be recovered by this project.[30] Several factors contributed to the success of the project: (1) The project was begun early in the life of the reservoir; (2) the formation rock was continuous and homogeneous;

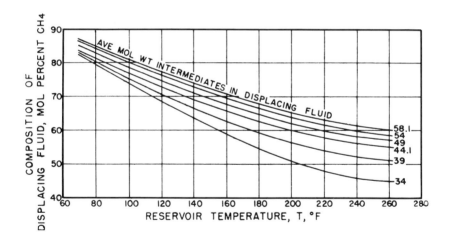

Fig. 6-13. Predicted phase conditions at 3000 psia for a miscible displacement of a reservoir fluid characterized by a C_{5+} molecular weight of 200. (After Benham et al.,[3] courtesy of the SPE of AIME.)

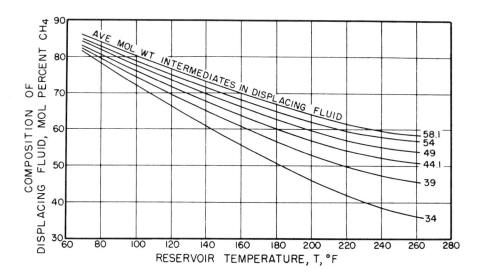

Fig. 6-14. Predicted phase conditions at 3000 psia for a miscible displacement of a reservoir fluid characterized by a C_{5+} molecular weight of 220. (After Benham et al.,[3] courtesy of the SPE of AIME.)

and (3) close engineering control over the project ensured miscible displacement and maximum sweep efficiency.[31]

In 1966, flue gas was substituted for natural gas after determining its ability to maintain miscibility. The flue gas is cheaper to use than hydrocarbon gas even on considering the amortization of a new flue gas generating plant. Besides economy, on using flue gas, more oil will be produced as the reservoir's economic limit will be lower.[31] References 31 to 33 provide detailed information on the use of flue gas in oil field operations.

A miscible displacement flood was begun in 1960 in the Wolfcamp Field, Texas.[34] Enriched gas in an amount equal to 2% of the total hydrocarbon pore volume was injected into the reservoir and was followed by a buffer of dry gas and then by alternate injections of water and gas as a driving medium. Water was injected to help improve the sweep efficiency.[35] At the time of the report (1963) the flood front had not reached the first wells; consequently, the efficiency could not be estimated. It was originally estimated that the field's 17% recovery under primary operations will be increased to about 50% under the miscible displacement flood.

In 1958, a miscible slug project was started in the San Andres Zone of the Slaughter Field, Texas.[15] An irregular-pattern flood was used. The LPG slug was injected first, followed by gas and, finally, by water. It was estimated that under primary operations, 20% of the oil-in-place would have been recovered, whereas the combined slug–gas–water process should result in 62% recovery. According to Sessions,[15] the project was progressing as planned, but owing to the short period of time since the start of the

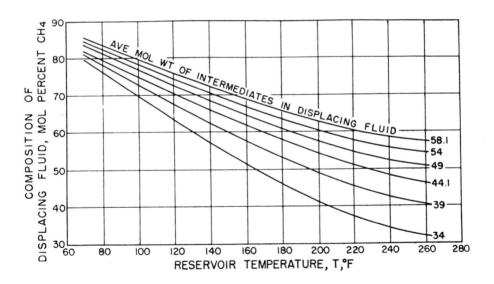

FIG. 6-15. Predicted phase conditions at 3000 psia for a miscible displacement of a reservoir fluid characterized by a C_{5+} molecular weight of 240. (After Benham et al.,[3] courtesy of the SPE of AIME.)

flood, the amount of data gathered was insufficient for full evaluation purposes.

In 1957, a miscible slug project was started in the Parks Field, Texas, in the Pennsylvanian Bend reservoir.[35] A slug of propane (4% of the total hydrocarbon pore volume) was injected followed by dry gas. In 1961, Marrs[35] estimated that the 17% recovery by primary means would be increased to 55%. There is also a five-spot pilot waterflood project underway in the field. Its recovery is estimated at 41% of the oil-in-place.

Fracture–Matrix Porosity System

There have been no reported miscible floods in carbonate reservoirs with a fracture–matrix porosity system. This is not too surprising, as the bypassing problems would probably be quite severe and the process would, therefore, be relatively inefficient.

Vugular–Solution Porosity System

There have been several miscible flood projects reported for carbonate reservoirs having a vugular–solution porosity system. All were slug processes with gas as the driving force. A miscible pilot flood was started in late 1956 in the Millican Field, Texas.[36] The volume of the propane–butane slug was equal to 1.5% of the floodable pore volume. The premature breakthrough of the gas was attributed to the small slug size and an unfavorable permeability distribution.[36]

In the Golden Spike Flood, Alberta, Canada, the total volume of the slug was 7% of the total hydrocarbon pore volume.[37] The data developed by Benham et al.[3] were used in determining the composition of the solvent bank. The suitability of the estimated composition was then confirmed by laboratory tests. Reservoir simulation was used to determine the optimum method for the placement of the slug. The ultimate recovery is expected to be near 96% of stock tank oil-in-place.[38] The gas injection project utilizing gravity-controlled rates of injection had been expected to recover 70%.[38]

The Wizard Lake D-3 project, Alberta, Canada, employs water injection into the aquifer as well as the miscible slug–gas cap injection system.[39] The water injection is required to move the oil–water contact up into the main body of the reservoir and hold it there because the aquifer is in pressure contact with other D-3 reservoirs in the area. Both oil and solvent could be lost to the aquifer if the oil–water contact control were not exercised. The flood is expected to raise the field's recovery from 66% for natural depletion (combination of water drive and gravity drainage) to 84%.

Only a single injector was used in the Millican Field pilot in Texas, and the sweep was horizontal.[36] In the floods of the three fields located in Alberta, Canada (Wizard Lake D-3A,[39] Golden Spike,[37] and the Rainbow Keg River[40]), the sweep was vertical. In the latter three projects, using one or more injectors, a slug of solvent was placed or generated at the gas–oil contact and the produced gas was injected into the gas cap to drive the slug vertically downward. Care was taken to maintain the rate of advance below the critical rate for the reservoir. Fingering and by-passing is thereby avoided as the gravity forces are permitted to control the process. The size of the slug required for the above conditions may be determined as follows[37]:

$$L = 3.62 \left(\frac{Dt}{\phi F}\right)^{1/2} \tag{6-2}$$

where:

L = mixed zone length, ft.
D = molecular diffusion coefficient, ft^2/day.
F = formation resistivity factor (determined by log analysis).
ϕ = porosity, fraction.
t = time for mixing (project life), days.

The Rainbow Keg River Field, Alberta, Canada, is composed of pinnacle reefs, which are being flooded using a miscible slug flood. The slug is generated in the reservoir at the gas–oil contact using enriched-gas injection,[40] a procedure which is also used in the Golden Spike Field.[37] The flood is expected to increase the recovery from 35% under primary depletion to about 95%.

Questions and Problems

6-1. Discuss the factors that affect the sweep efficiency of a miscible flood. Why would one anticipate sweep efficiencies to be lower for a miscible displacement in a massive limestone than for water?

6-2. List the methods which can be used to develop a miscible drive.

6-3. Describe how a miscible bank is formed by high-pressure gas injection.

6-4. Briefly describe the factors that govern the selection of an LPG slug size.

6-5. Calculate the size of the miscible slug required for a reef reservoir where t = 1 year, ϕ = 30%, F = 1.5, D = 0.009 ft^2/day.

References

1. Enright, R. J.: "Are Miscible Floods Worth the Cost?" *Oil and Gas J.* (Dec. 11, 1961) 43–46.
2. Clark, N. J., Shearin, H. M., Schultz, W. P., Garms, K. and Moore, J. L.: "Miscible Drive—Its Theory and Application", *J. Pet. Tech.* (June, 1958) 11–23.
3. Benham, A. L., Dowden, W. E. and Kunzman, W. J.: "Miscible Fluid Displacement—Prediction of Miscibility", *Trans.*, AIME (1960) **219**, 229–237.
4. Crosby, G. E.: "Is Miscible Flooding Dead?" *Pet. Eng.* (April, 1969) 53–55.
5. Scott, E. Z. and Read, D. L.: "A Study of Variables in Linear Miscible Displacement" SPE paper 1377-G presented at SPE Regional Meeting in Pasadena, California (Oct. 22–23, 1959).
6. (Editorial) "Miscible Flood May Capture 95 Percent of Crude In Place", *Oil and Gas J.* (July 26, 1965) 164–166.
7. Mahaffey, J. L., Rutherford, W. M. and Matthews, C. S.: "Sweep Efficiency by Miscible Displacement in a Five-Spot", *Soc. Pet. Eng. J.* (March, 1966) 73–80.
8. Habermann, B.: "The Efficiency of Miscible Displacement as a Function of Mobility Ratio", *Trans.*, AIME (1960) **219**, 264–272A.
9. Slobod, R. L. and Koch, H. A.: "High Pressure Gas Injection Mechanism of Recovery Increase", *Drill. and Prod. Pract.*, API (1953) 82–94.
10. Smith, C. R.: *Mechanics of Secondary Recovery*, Reinhold, New York (1966).
11. Welge, H. J., Johnson, E. F., Ewing, S. P. and Brinkman, F. H.: "The Linear Displacement of Oil from Porous Media by Enriched Gas", *J. Pet. Tech.* (Aug., 1961) 707–796.
12. Arnold, C. W., Stone, H. L. and Luffel, D. L.: "Displacement of Oil by Rich-Gas Banks", *Trans.*, AIME (1960) **219**, 305–312.
13. Craig, F. F. and Owens, W. W.: "Miscible Slug Flooding—A Review", *J. Pet. Tech.* (April, 1960) 11–16.
14. Koch, H. A. and Slobod, R. L.: "Miscible Slug Process", *Trans.*, AIME (1957) **210**, 40–47.
15. Sessions, R. E.: "How Atlantic Operates the Slaughter Flood", *Oil and Gas J.* (July 4, 1960) 91–98.
16. Garder, A. O., Jr., Peaceman, D. W. and Pozzi, A. L., Jr.: "Numerical Calculation of Multidimensional Miscible Displacement by the Method of Characteristics", *Soc. Pet. Eng. J.* (March, 1964) 26–36.
17. Gardner, G. H., Downie, J. and Kendall, H. A.: "Gravity Segregation of Miscible Fluids in Linear Models", *Soc. Pet. Eng. J.* (June, 1962) 95–104.
18. Greenkorn, R. A., Johnson, C. A. and Haring, R. E.: "Miscible Displacement in a Controlled Natural System", *J. Pet. Tech.* (Nov., 1965) 1329–1335.
19. Peaceman, D. W. and Rachford, H. H., Jr.: "Numerical Calculation of Multidimensional Miscible Displacement", *Soc. Pet. Eng. J.* (Dec., 1962) 327–346.
20. Peaceman, D. W.: "Improved Treatment of Dispersion in Numerical Calculation of Multidimensional Miscible Displacement", *Soc. Pet. Eng. J.* (Sept., 1966) 213–216.
21. Koval, E. J.: "A Method for Predicting the Performance of Unstable Miscible Displacement in Heterogeneous Media", *Soc. Pet. Eng. J.* (June, 1963) 145–154.
22. Wilson, J. F.: "Miscible Displacement—Flow Behavior and Phase Relationships for a Partially Depleted Reservoir", *Trans.*, AIME (1960) **219**, 223–228.
23. Rowe, A. M., Jr. and Silberberg, I. H.: "Prediction of the Phase Behavior Generated by the Enriched-Gas-Drive Process", *Soc. Pet. Eng. J.* (June, 1965) 160–166.

24. Doepel, G. W. and Sibley, W. P.: "Miscible Displacement—A Multilayer Technique for Predicting Reservoir Performance", *J. Pet. Tech.* (Jan., 1962) 73–80.
25. Agan, J. B. and Fernandes, R. J.: "Performance of a Miscible Slug Process in a Highly Stratified Reservoir", *J. Pet. Tech.* (Jan., 1962) 81–86.
26. Coats, K. H.: "Use and Misuse of Reservoir Simulation Models", *J. Pet. Tech.* (Nov., 1969) 1391–1398.
27. McCulloch, R. C., Langton, J. R. and Spivak, A.: "Stimulation of High Relief Reservoirs, Rainbow Field, Alberta, Canada", *J. Pet. Tech.* (Nov., 1969) 1399–1408.
28. Fitch, R. A. and Griffith, J. D.: "Experimental and Calculated Performance of Miscible Floods in Stratified Reservoirs", *J. Pet. Tech.* (Nov., 1964) 1209–1298.
29. Bleakley, W. B.: "Miscible Flood Hikes Block 31's Oil Output", *Oil and Gas J.* (Oct. 27, 1969) 67–70.
30. Herbeck, E. F. and Blanton, J. R.: "Ten Years of Miscible Displacement in Block 31 Field", *J. Pet. Tech.* (June, 1963) 543–549.
31. Davison, K.: "Inert Gas Boasts Recovery from Heavy Oil Reservoirs", *World Oil* (March, 1967) 98–105.
32. Clark, N. J., Roberts, T. G. and Lindner, J. D.: "Engine Exhaust Gas Boasts Heavy Oil Recovery", *Pet. Eng.* (Aug., 1964) 43–47.
33. Tittle, R. M. and From, K. T.: "Success of Flue Gas Program at Neale Field", SPE preprint 1907 presented at the 42nd Annual Fall Meeting of the SPE of AIME in Houston, Texas (Oct. 1–5, 1967).
34. Fitch, R. A. and Holloway, H. D.: "Performance of Miscible Flood with Alternate Gas-Water Displacement in the Midland Farms Wolfcamp Field, Andrews County, Texas", SPE paper 626 presented at SPE Regional Meeting in New Orleans, Louisiana (Oct. 24–25, 1963).
35. Marrs, D. G.: "Field Results of Miscible-Displacement Program Using Liquid Propane Driven by Gas, Parks Field Unit, Midland County, Texas", *J. Pet. Tech.* (April, 1961) 327–332.
36. Sturdivant, W. C.: "Pilot Propane Project Completed in West Texas Reef", *J. Pet. Tech.* (May, 1959) 27–30.
37. Larson, V. C., Peterson, R. B. and Lacey, J. W.: "Technology's Role in Alberta's Golden Spike Miscible Project", PDDC-12, *Seventh World Pet. Cong.*, Elsevier Publ. Co. (1967) **3**, 533–544.
38. Willman, G. J.: "Vertical Miscible Flood to Hike Recovery by 70 Million Barrels", *World Oil* (Jan., 1966) 75–78.
39. (Editorial) "Texaco Miscible Flood Program Begins for Alberta Pool "Soon"", *Oil and Gas J.* (Dec. 22, 1969) 48–49.
40. (Editorial) "Rainbow Project May Hit 96 Percent Recovery", *Oil and Gas J.* (Dec. 16, 1968) 48–49.

APPENDIX A

Determination of Volumes and Average Heights of Fractures

Introduction

Laboratory measured values of permeability for carbonate reservoir rocks are often significantly lower than values determined by well pressure buildup analysis and flow tests. The difference most likely is caused by the presence of fractures and solution cavities which are not adequately sampled in the cores analyzed. Whereas these fractures and solution channels may not contain a significant volume of oil, generally less than 5% of the total oil in a reservoir, they are very important to the attainment of economic production rates. Determination of the size of fractures in the laboratory is practically impossible, because the cores containing fractures of practical significance are commonly lost in the process of recovery. In addition, some fractures form during the recovery of cores.

On considering the impact of a fracture system on reservoir performance, Muskat[1] stated that the treatment of a general fractured limestone system, containing a number of fractures distributed at random, is not as yet practical mathematically. The significance of the fractures as fluid carriers can, however, be evaluated by considering a single fracture extending for some distance into the body of rock and opening into the wellbore.[1] The fluid-carrying capacity of this fracture can be determined on using the classical hydrodynamics equation for narrow linear channels (Eq. A-1). Muskat[1] also calculated the steady-state homogeneous-fluid flow characteristics of the limestone-fracture systems as a function of the fracture width. As shown in Fig. A-1, fracture widening can explain any increase of production capacity, following acid treatment, for example.[1] A given increase in fracture width will result in a greater relative increase in production capacity for formations with smaller production rates and initial fracture widths. The effect of fracture widening is also more pronounced in the case of lower-permeability carbonate rocks (dashed curves, Fig. A-1).

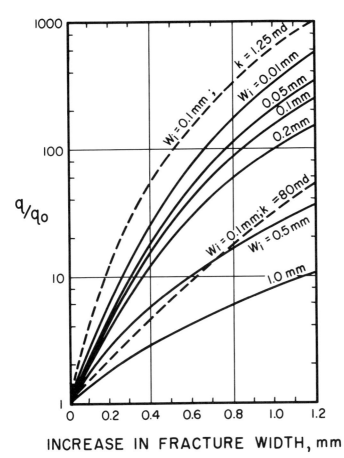

Fig. A-1. The calculated increase in the steady-state homogeneous-fluid production capacity of a fractured carbonate rock owing to increase in fracture width; q_o = production capacity before treatment, q = production capacity after treatment, w_i = initial fracture width. For solid curves, permeability of carbonates $k = 10$ md. (After Muskat,[1] p. 253, courtesy of McGraw-Hill Book Co.)

Derivation of Tank Oil-in-Place in Fractured Reservoirs[2,4]

On assuming that there is one rectangular fracture per cm² of cross-sectional area, it is possible to derive the tank oil-in-place, V_o, as follows starting with the classical hydrodynamics equation for narrow linear channels[2]:

$$q = \frac{b^3 a \, \Delta p}{12 \mu L} \tag{A-1}$$

where:

q = volumetric rate of flow, cm³/sec;
b = height or thickness of fracture, cm;
a = width of fracture, cm $(a \gg b)$;
L = length of fracture, cm;
Δp = pressure drop, dynes/cm²;
μ = viscosity, dyne-sec/cm².

Velocity of fluid in the fracture is thus equal to

$$v = q/ba = \frac{b^2 \Delta p}{12\mu L} \qquad (A\text{-}2)$$

and velocity per unit of cross-sectional area, A, in the rock is

$$v_r = \frac{b^2 \Delta p \phi}{12\mu L} \qquad (A\text{-}3)$$

where ϕ is fractional porosity and is equal to ba/A. Thus

$$b = \frac{\phi A}{a} \qquad (A\text{-}4)$$

Substitution of Eq. A-4 for b in Eq. A-3 gives

$$v_r = \frac{\phi^3 A^2 \Delta p}{12\mu L a^2} \qquad (A\text{-}5)$$

On assuming that there is one fracture per cm² of area, $A = 1$ cm² and $a = 1$ cm; thus

$$v_r = \frac{\phi^3 \Delta p}{12\mu L} \qquad (A\text{-}6)$$

and for radial flow

$$q = \frac{\pi h \phi^3 \Delta p}{6\mu B \ln(r_e/r_w)} \qquad (A\text{-}7)$$

where:

h = thickness of formation, cm;
r_e = drainage radius, cm;
r_w = wellbore radius, cm;
B = formation volume factor, bbl/STB.

Equation A-7 can be expressed as

$$q = 1.93 \times 10^8 \frac{h \phi^3 \Delta p}{\mu B \log(r_e/r_w)} \qquad (A\text{-}8)$$

if h is in m, μ in cp, q in m³/day, and Δp in atm. The productivity index in m³/day/atm is thus equal to

$$J = 1.93 \times 10^8 \frac{h\phi^3}{\mu B \log(r_e/r_w)} \tag{A-9}$$

Solving Eq. A-9 for porosity yields

$$\phi = 0.00173 \sqrt[3]{\frac{JB\mu \log(r_e/r_w)}{h}} \tag{A-10}$$

Inasmuch as porosity varies over short distances in carbonate reservoirs, it is necessary to divide the total productive surface area into smaller areas A_1, A_2, A_3, ..., having thicknesses h_1, h_2, h_3, ..., and porosities ϕ_1, ϕ_2, ϕ_3, ... Thus, the average fractional porosity is equal to

$$\phi_a = \frac{\phi_1 A_1 h_1 + \phi_2 A_2 h_2 + \phi_3 A_3 h_3 + \cdots}{A_1 h_1 + A_2 h_2 + A_3 h_3 + \cdots} \tag{A-11}$$

The volume of oil in fractures is equal to $\phi_a \times$ volume of productive formation. If the oil is present not only in fractures but also in pores, then the volume of oil is equal to

$$V_o = A_s h [\phi_p (1 - S_{wp})(1 - \phi_f) + \phi_f (1 - S_{wf})](1/B) \tag{A-12}$$

where V_o is in m³, A_s = surface area of bed in m², h = average thickness of bed in m, ϕ_p = total fractional porosity without fractures, S_{wp} = water saturation in pores, S_{wf} = water saturation in fractures, and ϕ_f = fractional porosity owing to fractures only.

On considering recovery factors,

$$V_o = \frac{A_s h}{B} [\phi_p (1 - S_{wp})(1 - \phi_f) E_p + \phi_f (1 - S_{wf}) E_f] \tag{A-13}$$

where:

E_p = recovery factor for porous portion of the formation
E_f = recovery factor for the fractured portion of the formation.

In the absence of pores

$$V_o = \frac{A_s h}{B} \phi_f (1 - S_{wf}) E_f \tag{A-14}$$

Kotyakhov, Serebrennikov, and Shcherbakova (*in* Khanin,[3] p. 66) presented the following formula for determining fracture porosity:

$$\phi_f = \frac{1}{577.9} \sqrt[3]{\frac{q_f B \mu \log(r_e/r_w) f_s^2}{h}} \tag{A-15}$$

where:

- ϕ_f = fractional fracture porosity;
- q_f = volumetric rate of flow from fractures, m³/day;
- B = formation volume factor;
- μ = dynamic viscosity, cp;
- r_e = wellbore radius, m;
- h = formation thickness, m;
- f_s = specific density of fractures (ratio of total fracture lengths in photographs to total area of photographs), as determined from thin sections or photographs of wellbore walls obtained at depth, cm/cm² or 1/cm.

Pozinenko (*in* Khanin,[3] p. 67) developed the following formula for determining fracture porosity of oil reservoir rocks, taking in consideration their anisotropy:

$$\phi_f = \frac{1}{577.9} \sqrt[3]{\frac{BJ\mu \log(r_e/r_w)f_d^2}{h}\left[\left(\frac{k_{11}}{k_{22}}\right)^{1/2} + \left(\frac{k_{22}}{k_{11}}\right)^{1/2}\right]} \quad \text{(A-16)}$$

where:

- ϕ_f = fractional fracture porosity;
- B = formation volume factor;
- J = productivity index, m³/day;
- μ = viscosity, cp;
- r_e = drainage radius, m;
- r_w = wellbore radius, m;
- f_d = volumetric density of open fractures (ratio of area of fractures to volume), 1/cm;
- h = thickness of formation, m;
- k_{11} and k_{22} are extreme permeabilities of the formation, d.

On the other hand, the fracture porosity for gas reservoir rocks can be determined from the following formula:

$$\phi_f = \frac{1}{458} \sqrt[3]{\frac{(273 + T_f)Z\mu \log(r_e/r_w)f_d^2}{(273 + T_s)h} \times \frac{q_g}{p_e^2 - p_w^2}\left[\left(\frac{k_{11}}{k_{22}}\right)^{1/2} + \left(\frac{k_{22}}{k_{11}}\right)^{1/2}\right]}$$

(A-17)

where:

- T_f = formation temperature, °C;
- T_s = surface temperature, °C;
- z = compressibility factor;
- μ = viscosity of gas, cp;

Determination of Volumes and Average Heights of Fractures

q_g = volume of gas production, m³/day;
p_e = static pressure, atm;
p_w = flowing pressure, atm;
f_d = volumetric density of fractures determined from the formula $f_d = (\pi/2)(l/A)$, where l = lengths of fractures in thin section, A = area of thin section (f_d can also be determined visually from cores).

Determination of Average Height of Fractures

The average height of fracture can be derived as follows, assuming the flow occurs only through the fractures.[4]

The volumetric rate of flow, q, in cm³/sec, is equal to

$$q = \frac{b^3 a \Delta p}{12\mu L} = \frac{kA\Delta p}{\mu L} \tag{A-18}$$

where:

b = height of fracture in cm;
a = width of fracture in cm;
L = length of fracture in cm;
μ = poise;
Δp = dynes/cm².

Therefore,

$$k = \frac{b^3 a}{12A} \tag{A-19}$$

but

$$\phi = \frac{ba}{A} \quad \text{and} \quad a = \frac{\phi A}{b} \tag{A-20}$$

on substituting Eq. A-20 in Eq. A-19:

$$k = \frac{\phi b^2}{12} \tag{A-21}$$

and if k is in darcys, then

$$k = 83 \times 10^5 \phi b^2 \tag{A-22}$$

Thus, the height of fracture, b, in cm, can be calculated:

$$b = \left(\frac{k}{83 \times 10^5 \phi}\right)^{1/2} \tag{A-23}$$

Sample Problem[2]

Calculate ϕ, k, and b (height or thickness of fracture) if $J = 10$ m^3/day/atm, $\mu = 10$ cp, $r_e = 500$ m, $r_w = 0.2$ m, $h = 10$ m, and $B = 1.3$.

Solution:

Use Eq. A-10 to get porosity

$$\phi = \frac{1}{577.9} \sqrt[3]{\frac{10 \times 1.3 \times 10 \times 3.4}{10}} = 0.0061 \text{ or } 0.61\%$$

In order to determine permeability, the following formula can be used:

$$k = \frac{JB\mu \log(r_e/r_w)}{23.6h} = \frac{10 \times 1.3 \times 10 \times 3.4}{23.6 \times 10} = 1.87 \text{ darcys}$$

Substitution of the values for ϕ and k in Eq. A-23 gives

$$b = \left(\frac{1.87}{83 \times 10^5 \times 0.0061}\right)^{1/2} = 0.0061 \text{ cm or } 61 \text{ microns}$$

If $J = 1$ m^3/day/atm, then $\phi = 0.28\%$, $k = 0.187$ d, and $b = 28$ microns.

References

1. Muskat, M.: *Physical Principles of Oil Production*, 1st ed., McGraw-Hill, New York (1949) 922 pp.
2. Kotyakhov, F. I.: "Approximate Method of Determining Petroleum Reserves in Fractured Rocks", *Neftyanoe Khozyaystvo* (April, 1956) **4**, 40–46.
3. Khanin, A. A.: *Oil and Gas Reservoir Rocks and Their Study*, Izd. "Nedra", Moscow (1969) 366 pp.
4. Chilingar, G. V.: "Approximate Method of Determining Reserves and Average Height of Fractures in Fractured Rocks: An Interim Report", *Compass of Sigma Gamma Epsilon* (1959), **36**, 202–205.

APPENDIX B

Fundamentals of Surface and Capillary Forces

Introduction

Wettability may be defined as the ability of the liquid to wet, or spread over, a solid surface. Figure B-1,A shows a liquid wetting a solid. On the other hand, Fig. B-1,B shows the relationship between the liquid and solid, when the liquid has little affinity for the solid. In Fig. B-1,C the liquid drop occupies an intermediate position. The fluid which wets the surface more strongly, occupies the smaller pores and minute interstices.

Interfacial Tension and Contact Angle[1,2,3]

The angle which the interface makes with the solid is called the contact angle, θ. Usually, it is measured from the solid through the liquid phase (if the other phase is a gas) and through the water phase if oil and water are both present. In a capillary tube, shown in Fig. B-2,A, the angle between the side of the tube and the tangent to the curved interface (where it intersects the side of tube) is less than 90°. For a capillary depression, shown in Fig. B-2,B, the contact angle is greater than 90°. In the case of no rise or depression, the angle is 90° (Fig. B-2,C).

Interfacial tension, σ, is caused by the molecular property (intermolecular cohesive forces) of liquids. It has the dimensions of force per unit length (lb/ft or dynes/cm), or energy per unit area (ergs/cm²). On considering an element of a surface having double curvature (r_1 and r_2), the sum of the force components normal to the element is equal to zero (Fig. B-3). The pressure difference, $p_2 - p_1$, is balanced by the interfacial tension forces:

$$(p_2 - p_1)dydx = 2\sigma dy \sin \theta_2 + 2\sigma dx \sin \theta_1 \tag{B-1}$$

If the contact angles θ_1 and θ_2 are small, the following simplifications

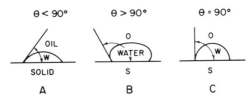

Fig. B-1. Different degree of wetting of solid by a liquid.

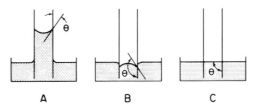

Fig. B-2. Behavior of various fluids in glass capillary tubes. A, water; B, mercury; and C, tretrahydronaphthalene (when glass is perfectly clean and liquid is pure).

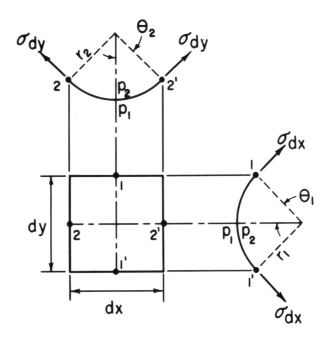

Fig. B-3. Surface tension forces acting on a small element on the surface having double curvature.[1,2] $(p_2 = p_1 + \gamma h)$

Fundamentals of Surface and Capillary Forces

may be made:

$$\sin \theta_1 = \frac{dy}{2r_1} \tag{B-2}$$

and

$$\sin \theta_2 = \frac{dx}{2r_2} \tag{B-3}$$

Therefore, Eq. B-1 becomes:

$$p_2 - p_1 = \sigma \left(\frac{1}{r_2} + \frac{1}{r_1}\right) \tag{B-4}$$

For a capillary tube (Fig. B-4):

$$r_1 = r_2 = r \tag{B-5}$$

$$\cos \theta = \frac{d}{2r} \tag{B-6}$$

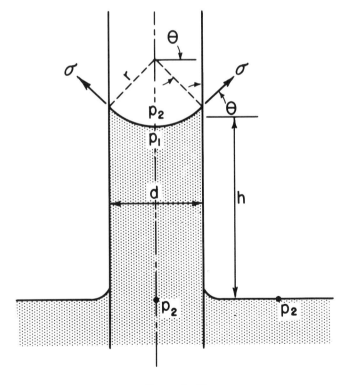

FIG. B-4. Rise of water in a glass capillary tube.[1,2]

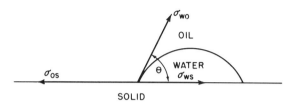

FIG. B-5. Shape of water drop resulting from interfacial tension forces.

and

$$p_2 = p_1 + \gamma h \tag{B-7}$$

where γ = specific weight of fluid, d = diameter of capillary tube, and h = height of capillary rise. Thus, Eqs. B-4 through B-7 may be combined to yield the following expression for capillary rise, h:

$$h = \frac{4\sigma \cos \theta}{\gamma d} \tag{B-8}$$

Equation B-8 can also be derived on considering the equilibrium of vertical forces. The weight of fluid in the capillary tube, W, which is acting downward, is equal to

$$W = \frac{\pi}{4} d^2 h \gamma \tag{B-9}$$

The vertical component of interfacial tension force acting upwards is equal to:

$$F_{\sigma y} = \pi d \sigma \cos \theta \tag{B-10}$$

Equating these two forces and solving for h give rise to Eq. B-8.

In reference to Fig. B-5, the interfacial tensions can be expressed as

$$\sigma_{ws} + \sigma_{wo} \cos \theta = \sigma_{so} \tag{B-11}$$

where σ_{ws}, σ_{wo}, and σ_{so} = interfacial tensions at the phase boundaries water–solid, water–oil and solid–oil, respectively, or

$$\cos \theta = \frac{\sigma_{so} - \sigma_{ws}}{\sigma_{wo}} \tag{B-12}$$

As shown in Fig. B-6,A, when solid is completely immersed in water phase, $\theta = 0°$, $\cos \theta = 1$, and consequently

$$\sigma_{wo} = \sigma_{so} - \sigma_{ws} \tag{B-13}$$

When half of the solid is wet by water and the other half by oil (Fig. B-6,B), $\theta = 90°$, $\cos \theta = 1$ and thus

$$\sigma_{so} = \sigma_{ws} \qquad \text{(B-14)}$$

On the other hand, if solid is completely wet by oil (Fig. B-6,C), $\theta = 180°$, $\cos \theta = -1$, and

$$\sigma_{so} = \sigma_{ws} - \sigma_{wo} \qquad \text{(B-15)}$$

If $\theta < 90°$, the surfaces are called hydrophilic and when $\theta > 90°$ they are called hydrophobic. An interfacial-tension depressant lowers σ_{wo}, whereas a wetting agent lowers θ or increases $\cos \theta$. A decrease in σ_{wo} does not necessarily mean an increase in $\cos \theta$, or vice versa, because of the changes in σ_{so} and σ_{ws}.

If a rock is completely water wet ($\theta = 0$), water will try to envelop all of the grains and force all of the oil out in the middle of the pore channel. Even though some oil may still be trapped in this case, the recovery would be high. On the other hand, if all of the solid surfaces were completely oil wet ($\theta = 180°$), oil would try to envelop all of the grains and force all of the water out into the center of the pore channel. In this extreme case, recovery would be very low by water drive. Some oil-wet reservoirs are known to exist.

In the usual case ($0° < \theta < 180°$), to improve waterflooding operations the contact angle θ should be changed from $>90°$ to $<90°$, through the use of surfactants. This would move the oil from the surface of the grains out into the center of the pore channels.

Contaminants or impurities may exist in either fluid phase and/or may be adsorbed on the solid surface. Even if present in minute quantities, they can and do change the contact angle from the value measured for pure systems.[3]

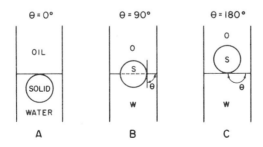

FIG. B-6. Illustrations of 0°, 90°, and 180° contact angles.

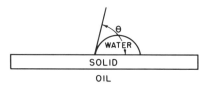

Fig. B-7. Contact angle: plate first immersed in oil followed by the placement of water drop on top.

Effect of Contact Angle and Interfacial Tension on Movement of Oil

For an ideal system composed of pure liquids, the advancing contact angle should equal the receding angle. Because of the presence of impurities within the liquids, however, the advancing contact angle is greater in most systems. The advancing angle of contact is the angle formed at the phase boundary when oil is displaced by water. It can be measured as follows: The crystal plate is covered by oil and then the water drop is advanced on it. The contact angle is the limiting angle with time after equilibrium has been established (Fig. B-7). The contact angle formed when water is displaced by oil is called receding angle (Fig. B-8). The contact angles during the movement of a water–oil interface in a cylindrical capillary having a hydrophilic surface are shown in Fig. B-9.

Inasmuch as a reservoir is basically a complex system of interconnected capillaries of various sizes and shapes, an understanding of flow through capillaries is very important. In Fig. B-10, a simple two-branch capillary system is presented. If a pressure drop is applied, the water will flow more readily through the large-diameter capillary than it will through the small-diameter one. Thus, a certain volume of oil may be trapped in the small capillary when the water reaches the upstream fork. Poiseuille's law states that

$$q = \frac{\pi d^4 \, \Delta p_t}{128 \mu L} \tag{B-16}$$

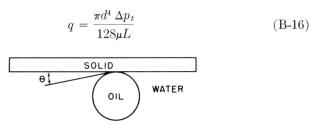

Fig. B-8. Contact angle: plate first immersed in water followed by placing a drop of oil underneath.

Fundamentals of Surface and Capillary Forces

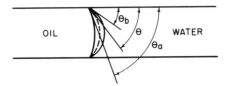

FIG. B-9. Changes in contact angle as a result of movement of water–oil interface. θ = contact angle at static position; θ_a = contact angle when oil is displaced by water (advancing angle); and θ_b = contact angle when water is displaced by oil (receding angle).

and

$$v = \frac{q}{A} = \frac{d^2 \, \Delta p_t}{32 \mu L} \tag{B-17}$$

where:

q = volumetric rate of flow, cm³/sec;
d = diameter of capillary, cm;
Δp_t = total pressure drop, dynes/cm²;
A = cross-sectional area, cm²;
μ = viscosity, cp;
L = flow path length, cm;
v = velocity, cm/sec.

The capillary pressure, P_c, is equal to

$$P_c = \frac{4\sigma \cos \theta}{d} \tag{B-18}$$

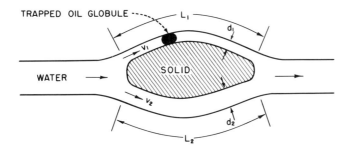

FIG. B-10. Flow through a two-branch capillary and trapping of oil in a small-diameter capillary.

where:

  σ = interfacial tension between oil and water, dynes/cm;
  d = diameter of capillary, cm;
  θ = contact angle.

The total pressure drop, Δp_t, is equal to

$$\Delta p_t = \Delta p_i + P_c \tag{B-19}$$

where Δp_i = applied pressure, dynes/cm². Solving for v in each capillary, by combining Eqs. B-17, B-18, and B-19 gives

$$v_1 = \frac{d_1^2}{32\mu_1 L_1}\left(\Delta p_i + \frac{4\sigma \cos\theta}{d_1}\right) \tag{B-20}$$

and

$$v_2 = \frac{d_2^2}{32\mu_2 L_2}\left(\Delta p_i + \frac{4\sigma \cos\theta}{d_2}\right) \tag{B-21}$$

Setting $L_1 = L_2$ and $\mu_1 = \mu_2$, and dividing Eq. B-20 by Eq. B-21 gives the following relationship:

$$\frac{v_1}{v_2} = \frac{d_1^2\,\Delta p_i + 4\sigma \cos\theta d_1}{d_2^2\,\Delta p_i + 4\sigma \cos\theta d_2} \tag{B-22}$$

Therefore, when $\Delta p_i \gg P_c$,

$$\frac{v_1}{v_2} \approx \frac{d_1^2}{d_2^2} \tag{B-23}$$

and when $\Delta p_i \ll P_c$,

$$\frac{v_1}{v_2} \approx \frac{d_1}{d_2} \tag{B-24}$$

As shown in Fig. B-11, the sum of forces acting on the trapped oil globule may be expressed as

$$\sum F = F_1 + F_2 - F_3 \tag{B-25}$$

where:

$$\sum F = \frac{\pi d^2}{4}\Delta p_t$$

$$F_1 = \frac{\pi d^2}{4}\Delta p_i$$

$$F_2 = \pi d(\sigma_a \cos\theta_a)$$

$$F_3 = \pi d(\sigma_b \cos\theta_b)$$

Fundamentals of Surface and Capillary Forces

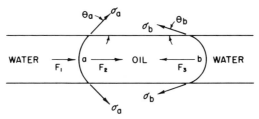

FIG. B-11. Forces acting on a trapped oil globule in a capillary.

thus,

$$\Delta p_t = \Delta p_i + \frac{4\sigma_a \cos \theta_a}{d} - \frac{4\sigma_b \cos \theta_b}{d} \quad \text{(B-26)}$$

Because the receding angle is usually less than the advancing angle, the capillary pressure not only does not help but also hinders the flow. The term $(4\sigma_b \cos \theta_b/d)$ is usually greater than $(4\sigma_a \cos \theta_a/d)$ because $\theta_b < \theta_a$. If a surfactant were added at the left to reduce σ_a, Δp_t would become less and the oil globule may eventually move to the left in Fig. B-11 when Δp_t becomes negative. The quantity of trapped oil is dependent upon the value of $\sigma \cos \theta$ at each end of the globule as well as upon Δp_i (imposed pressure drop).

Inasmuch as the contact angle depends upon the interfacial tensions, which, in turn, may be influenced by surfactants, these chemicals may alter recovery by altering both the contact angle and interfacial tension. As the oil is displaced by water, which wets the rock surfaces, capillary pressure is a driving force. If, on the other hand, water does not wet the rock surface, capillary pressure is a retarding force which must be overcome.

The magnitude of capillary pressure in pores having a radius of around 15 microns is not large and therefore capillary pressure is not an important force during the movement of oil–water contact, providing there is no mixing. The movement of oil and water in reservoir, however, results in the formation of water–oil and gas–water–oil mixtures.[4] The amount of gas coming out of solution during migration is greater with increasing amount of dissolved active substances, with increasing surface area of porous medium (i.e., with decreasing permeability), and with decreasing temperature. As the oil–water–gas mixtures move through pores, the gas bubbles and water droplets are deformed on passing through constrictions (Fig. B-12).[4] In order to move, the gas bubble shown in Fig. B-12 must overcome the capillary pressure equal to

$$\Delta p = p_1 - p_2 = \frac{2\sigma}{r_1} - \frac{2\sigma}{r_2} = 2\sigma \left(\frac{1}{r_1} - \frac{1}{r_2} \right) \quad \text{(B-27)}$$

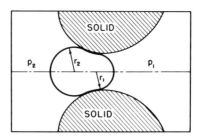

Fig. B-12. Movement of gas bubble through a constriction.[4]

Although the Δp may be very small for a single bubble, the cumulative resistance of many bubbles may be large (Jamin effect). Additional resistance to flow is created by the polymolecular layers of oriented molecules of surface-active components in the oil, which are adsorbed on the rock surface and may be quite thick (10^{-3} to 10^{-4} cm). At a constant pressure differential, the rate of oil filtration through porous media diminishes with time and is more pronounced in the case of higher content of polar components in the oil.

References

1. Binder, R. C.: *Fluid Mechanics*, 4th ed., Prentice-Hall, Englewood Cliffs, New Jersey (1962) 453 pp.
2. Vennard, J. K.: *Elementary Fluid Mechanics*, 4th ed., John Wiley, New York (1961) 570 pp.
3. Marsden, S. S.: "Wettability: The Elusive Key to Waterflooding", *Pet. Eng.* (April, 1968) 82–87.
4. Muravyov, I., Andriasov, R., Gimatudinov, Sh., Govorova, G. and Polozkov, V.: *Development and Exploitation of Oil and Gas Fields*, Peace (Mir) Publishers, Moscow (1958) 503 pp. (In English.)

APPENDIX C

Formation Volume Factors for Natural Gas, Water, and Crude Oils

Formation Volume Factor for Natural Gas

The formation volume factor for natural gas, B_g, in bbl/SCF may be calculated as follows, assuming $p_{sc} = 14.7$ psia and $T_{sc} = 60°F$ ($sc =$ standard conditions):

$$B_g = \frac{0.00504\, Tz}{p} \qquad \text{(C-1)}$$

where:

$z =$ gas deviation factor;
$T =$ reservoir temperature, °R;
$p =$ average reservoir pressure, psia.

If B_g is in cu ft/SCF, constant in Eq. C-1 is equal to 0.02829.

The gas compressibility factor, z, can be calculated using the work of Standing and Katz,[1] or it can be obtained from laboratory testing. The compressibility factor can be calculated as follows.

1. Using Fig. D-3, p. 200, obtain the pseudocritical pressure, p_c, and pseudocritical temperature, T_c.[2]
2. Calculate the pseudoreduced pressure, p_r, and pseudoreduced temperature, T_r:

$$p_r = \frac{\text{absolute pressure}}{p_c} \qquad \text{(C-2)}$$

$$T_r = \frac{\text{absolute temperature}}{T_c} \qquad \text{(C-3)}$$

3. Utilizing Fig. C-1, obtain the value of the gas deviation factor, z.[1]

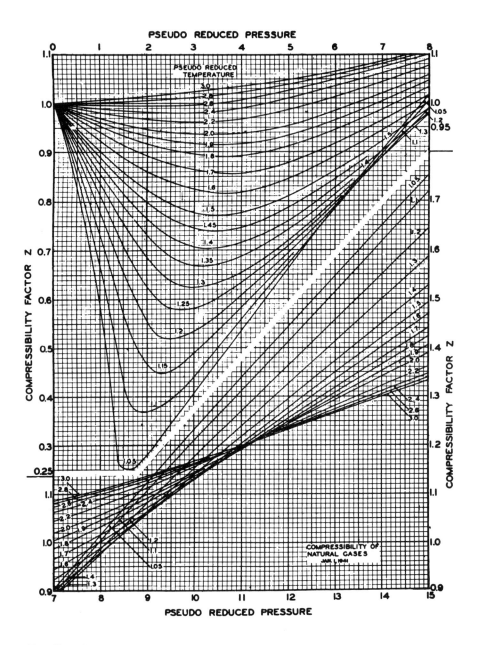

Fig. C-1. Compressibility factors for natural gases. (After Standing and Katz,[1] courtesy of AIME.)

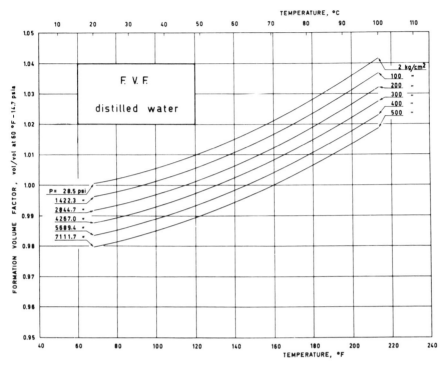

FIG. C-2. Formation volume factors of distilled water at various temperatures and pressures. (After Long and Chierici,[3] courtesy of *Petroleum Engineer*.)

Formation Volume Factor for Water

The formation volume factor for water may be obtained from Figs. C-2 through C-5.[3] A correction factor for dissolved gas may be obtained from Fig. C-6 and added to value obtained from Figs. C-2 through C-5.

Formation Volume Factor for Crude Oils

The formation volume factor for crude oils may be obtained from laboratory data or calculated on the basis of Standing's[2] and Lasater's[4] correlation charts.

Formation volume factors for crude oils for either differential or flash gas liberation are easily obtained in the laboratory. The formation volume

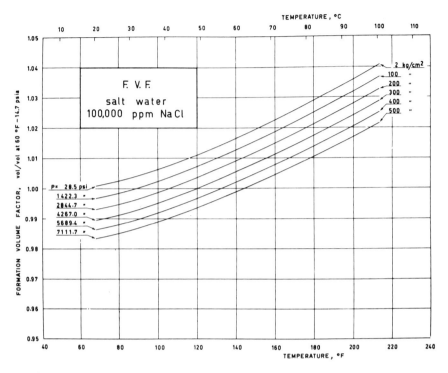

Fig. C-3. Formation volume factors of 100,000-ppm NaCl solution at various temperatures and pressures. (After Long and Chierici,[3] courtesy of *Petroleum Engineer*.)

factor desired is a combination of differential and flash factors[5]:

$$B_o = B_{od}\frac{B_{ofb}}{B_{odb}} \qquad (C\text{-}4)$$

where:

B_o = differential-flash factor, bbl/STB;
B_{odb} = differential formation volume factor at the bubble point pressure, bbl/STB;
B_{ofb} = flash formation volume factor at the bubble point pressure, bbl/STB;
B_{od} = differential formation volume factor at a given reservoir pressure, bbl/STB.

Empirical correlations to determine the bubble point pressure have been prepared for California (U.S.A.) crude oils[2] (Fig. C-7) and for crude oils from midcontinent (U.S.A.), Canada, and South America[4] (Fig. C-8). It is suggested that for California crude oils, Figs. C-9 and C-10 can be used directly. For midcontinent crudes, Amyx et al.[5] have suggested using the following equation based on the work of Katz[6]:

$$B_o = (1 + S_t)(1 + S_p) \qquad (C\text{-}5)$$

where S_t = fractional shrinkage of residual oil with respect to temperature of crude oil (Fig. C-11), and S_p = fractional shrinkage of residual oil due to the liberation of the solution gas (Fig. C-12).

For Venezuelan crude oils, Fig. C-13 can be used, and the formula for formation volume factor, B_o, is[7]:

$$B_o = B_R(1 + S_t) \qquad (C\text{-}6)$$

where B_R = thermally reduced oil formation volume factor, bbl/STB (Fig. C-13).

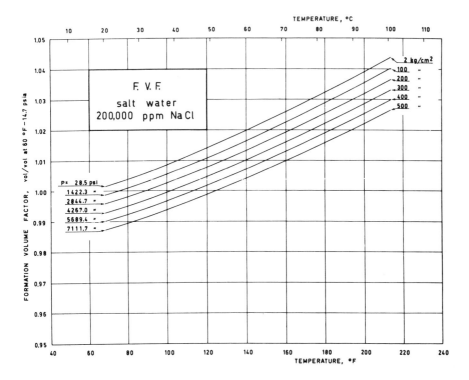

FIG. C-4. Formation volume factors of 200,000-ppm NaCl solution as a function of temperature and pressure. (After Long and Chierici,[3] courtesy of *Petroleum Engineer*.)

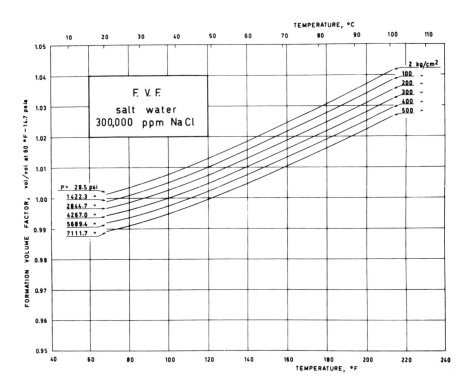

Fig. C-5. Formation volume factors of 300,000-ppm NaCl solution at various pressures and temperatures. (After Long and Chierici,[3] courtesy of *Petroleum Engineer*.)

Formation Volume Factors for Natural Gas, Water, and Crude Oils

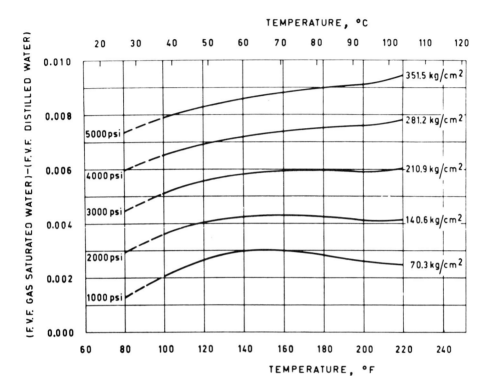

Fig. C-6. Effect of dissolved gas on formation volume factor of distilled water at various temperatures and pressures. (After Dodson and Standing, in: Long and Chierici,[3] courtesy of *Petroleum Engineer*.)

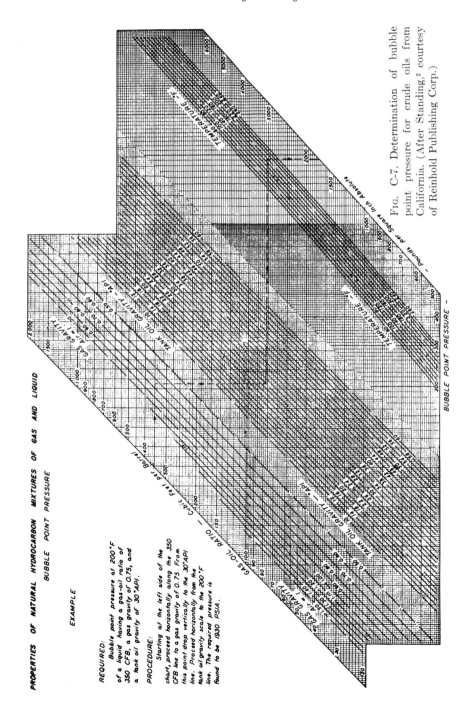

Fig. C-7. Determination of bubble point pressure for crude oils from California. (After Standing,[2] courtesy of Reinhold Publishing Corp.)

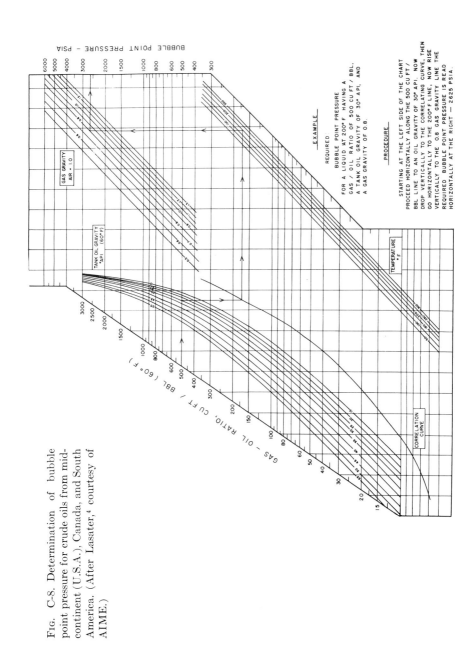

Fig. C-8. Determination of bubble point pressure for crude oils from mid-continent (U.S.A.), Canada, and South America. (After Lasater,[4] courtesy of AIME.)

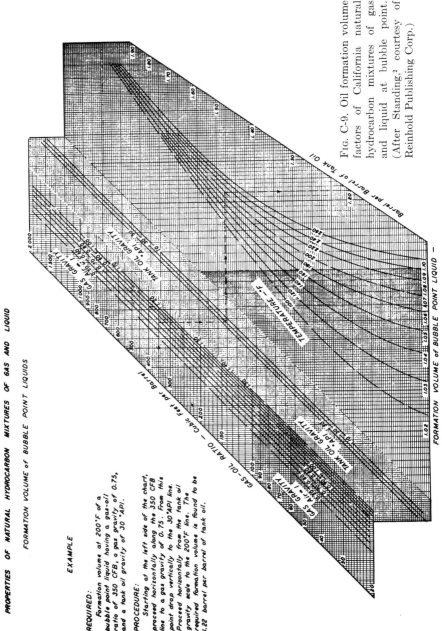

Fig. C-9. Oil formation volume factors of California natural hydrocarbon mixtures of gas and liquid at bubble point. (After Standing,[2] courtesy of Reinhold Publishing Corp.)

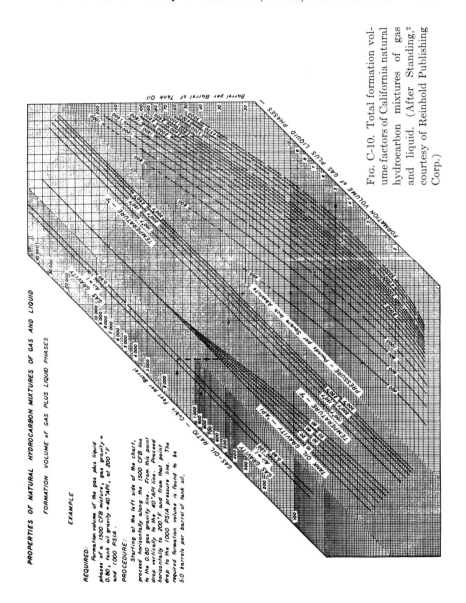

FIG. C-10. Total formation volume factors of California natural hydrocarbon mixtures of gas and liquid. (After Standing,[2] courtesy of Reinhold Publishing Corp.)

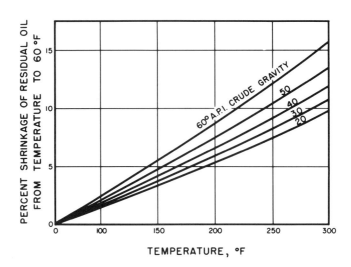

Fig. C-11. Fractional shrinkage of residual oil (expressed in %) with respect to temperature and crude oil gravity. (After Katz,[6] courtesy of API.)

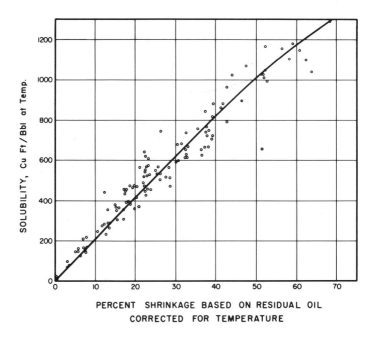

Fig. C-12. Fractional shrinkage of residual oil (expressed in %) due to the liberation of the solution gas. (After Katz,[6] courtesy of API.)

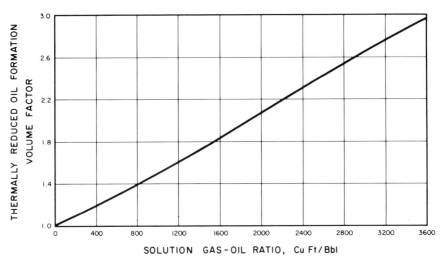

Fig. C-13. Thermally reduced oil formation volume factor versus solution gas–oil ratio for saturated crude oils. (After Knopp and Ramsey,[7] courtesy of the SPE of AIME.)

References

1. Standing, M. B. and Katz, D. L.: "Density of Natural Gases", *Trans.*, AIME (1942) **146,** 140.
2. Standing, M. B.: *Volumetric and Phase Behavior of Oil Field Hydrocarbon Systems*, Reinhold, New York (1952).
3. Long, G. and Chierici, G.: "Salt Content Changes Compressibility of Reservoir Brines", *Petroleum Engineer* (July, 1961) B-25–B-32.
4. Lasater, J. A.: "Bubble Point Pressure Correlation", *Trans.*, AIME (1958) **213,** 379–381.
5. Amyx, J. W., Bass, D. M. and Whiting, R. L.: *Petroleum Reservoir Engineering*, McGraw-Hill, New York (1960).
6. Katz, D. L.: "Prediction of the Shrinkage of Crude Oils", *Drill. and Prod. Pract.*, API (1942) 137–147.
7. Knopp, C. R. and Ramsey, L. A.: "Correlation of Oil Formation Volume Factor and Solution Gas-Oil Ratio", *J. Pet. Tech.* (Aug., 1960) 27–29.

APPENDIX D

Viscosities of Air, Water, Natural Gas, and Crude Oil

Introduction

The viscosities of air, water, natural gas, and crude oil are required for various mobility and fluid flow calculations. Various figures, which may be used if laboratory data for the project area are not available, are presented here.

Viscosity of Air

The viscosity of air as a function of temperature and pressure can be obtained from Fig. D-1.[1]

Viscosity of Water

Data on the viscosity of water was summarized by van Wingen[2] (Fig. D-2). These data have not been adjusted for the effect of dissolved gas; however, they can be used for a first approximation. If the success of the proposed project requires greater accuracy, then laboratory analyses using reservoir fluids should be made.

Viscosity of Natural Gas

The viscosity of natural gas has been studied by several investigators.[1,3] The following quick technique for determining the viscosity of a natural gas has been suggested by Carr et al.[3]:

1. Referring to Fig. D-3 determine the pseudocritical pressure, p_c, and pseudocritical temperature, T_c.

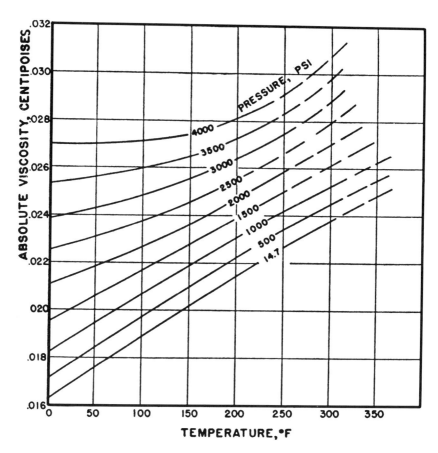

Fig. D-1. Viscosity of air at oil field temperatures and pressures. (After Beal,[1] courtesy of the SPE of AIME.)

2. Calculate the pseudoreduced pressure, p_r, and pseudoreduced temperature, T_r:

$$p_r = \frac{\text{absolute pressure}}{p_c} \quad \text{(D-1)}$$

$$T_r = \frac{\text{absolute temperature}}{T_c} \quad \text{(D-2)}$$

3. Utilizing Figs. D-4 and D-5 obtain the viscosity ratio, μ/μ_1, and viscosity, μ_1, at 14.7 psia or 1 atm pressure and at reservoir temperature.

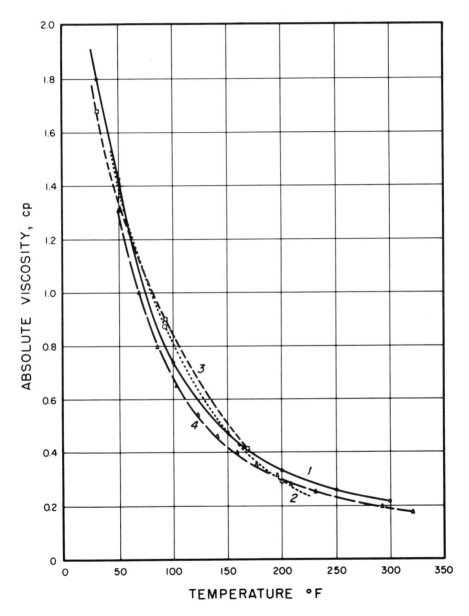

FIG. D-2. Relationship between viscosity of saline water (60,000 ppm) and temperature. Curve 1, at 14.7 psia; 2, at 14.2 psia; 3, at 7100 psia; 4, at vapor pressure. (After van Wingen,[2] courtesy of API.)

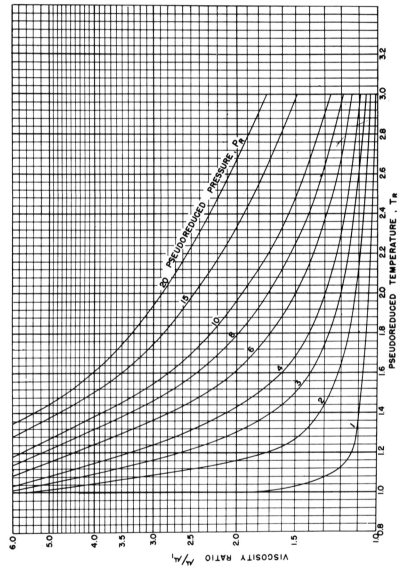

Fig. D-3. Viscosity ratio as a function of pseudoreduced temperature and pressure. (After Carr *et al.*,[3] courtesy of AIME.)

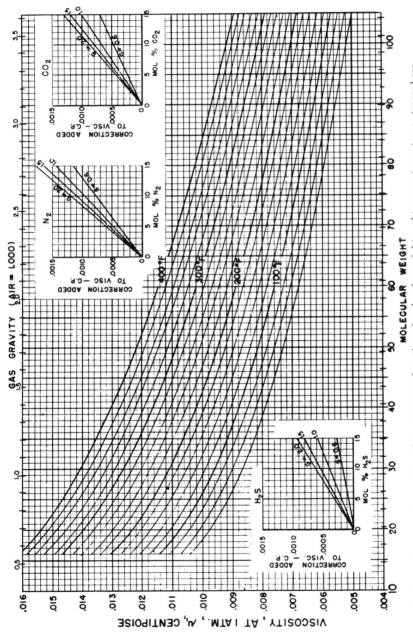

Fig. D-4. Viscosity of paraffin hydrocarbon gases as a function of molecular weight and temperature at one atmosphere. (After Carr et al.,[3] courtesy of AIME.) (G = gas gravity with respect to air.)

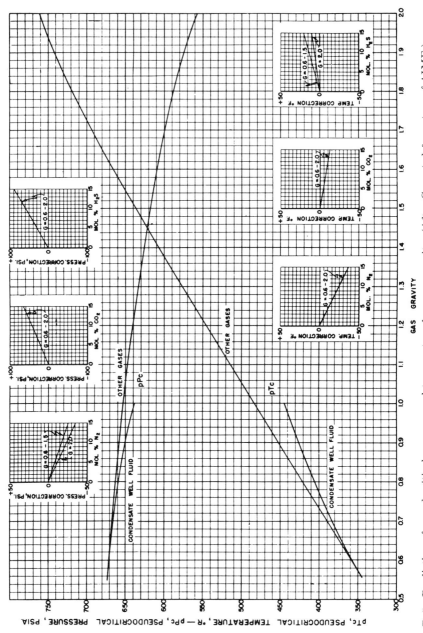

Fig. D-5. Prediction of pseudocritical pressure and temperature from gas gravity. (After Carr et al.,[3] courtesy of AIME.)

Viscosities of Air, Water, Natural Gas, and Crude Oil

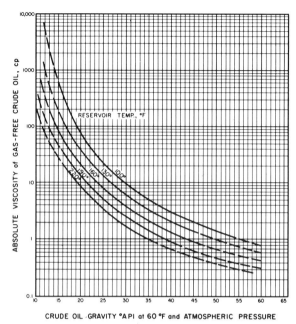

Fig. D-6. Viscosity of gas-free crude oils at oil field temperatures. (After Beal,[1] courtesy of the SPE of AIME.)

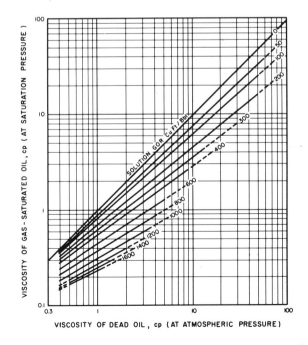

Fig. D-7. Viscosity of gas-saturated crude oils at saturation pressure. (After Chew,[4] courtesy of AIME.)

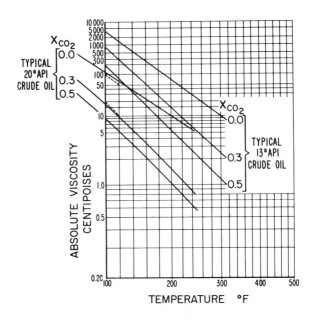

Fig. D-8. Relationship between viscosity and temperature for CO_2-crude oil mixtures. (After Simon and Graue,[5] courtesy of the SPE of AIME.) x_{CO_2} = mole fraction of CO_2 in a CO_2-crude oil mixtures.

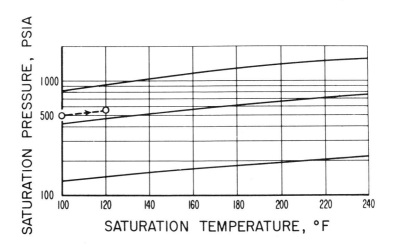

Fig. D-9. Relationship between saturation pressure of CO_2-crude oil mixtures and temperature. (After Simon and Graue,[5] courtesy of the SPE of AIME.)

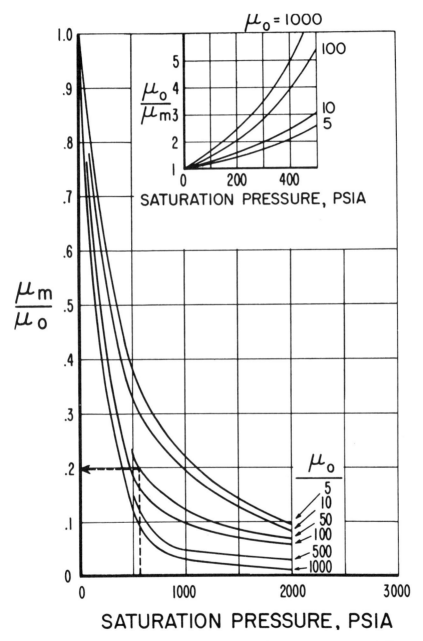

Fig. D-10. Viscosity of CO_2-crude oil mixtures at 120°F. (After Simon and Graue,[5] courtesy of the SPE of AIME.)

Note that molecular weight of gas = 28.95 × the specific gravity of gas with respect to air.

4. The viscosity at reservoir conditions, μ, is given by

$$\mu = (\mu/\mu_1)(\mu_1) \qquad (D-3)$$

Viscosity of Crude Oil

The viscosity for dead oil may be obtained from Fig. D-6. This value must be corrected for the dissolved gas content on using Fig. D-7.[4]

Simon and Graue have presented experimental data on solubility, swelling, and viscosity behavior of CO_2-crude oil systems.[5] Figures D-6 to D-10 are used as follows:

1. Determine the viscosity of the oil, μ_o, from Figs. D-6 and D-7 at the desired temperature. If specific data is not available, an approximate value may be obtained from Fig. D-8 using the line $x_{CO_2} = 0.0$ as a parallel guideline.
2. Determine the saturation pressure at the desired temperature by using the middle line in Fig. D-9 as a parallel guideline.
3. Directly read the ratio μ_m/μ_o from Fig. D-10. The insert in Fig. D-10 is intended to facilitate reading of the graph from 0 to 500 psia. The ordinate of the insert is the reciprocal of the ordinate of the main graph.
4. Calculate the viscosity of the mixture, μ_m:

$$\mu_m = \mu_o(\mu_m/\mu_o) \qquad (D-4)$$

5. Correct the viscosity of the mixture μ_m to other desired temperatures by use of Fig. D-8.

References

1. Beal, C.: "The Viscosity of Air, Water, Natural Gas, Crude Oil and Its Associated Gases at Oil Field Temperatures and Pressures", *Trans.*, AIME (1946) **165**, 94–115.
2. Van Wingen, N.: "Viscosity of Air, Water, Natural Gas and Crude Oil at Varying Pressures and Temperatures", in: *Secondary Recovery of Oil in the United States*, 2nd Ed., API (1950) 126–132.
3. Carr, N. L., Kobayashi, R. and Burrows, D. B.: "Viscosity of Hydrocarbon Gases Under Pressure", *Trans.*, AIME (1954) **201**, 264–272.
4. Chew, J.: "A Viscosity Correlation for Gas-Saturated Crude Oils", *Trans.*, AIME (1959) **216**, 23–25.
5. Simon, R. and Graue, D. J.: "Generalized Correlations for Predicting Solubility, Swelling and Viscosity Behavior of CO_2-Crude Oil Systems", *J. Pet. Tech.* (Jan., 1965) 102–106.

APPENDIX E

Rock and Fluid Compressibilities

Introduction to Estimating Reservoir Compressibilities

The total system compressibility, c_t, of the reservoir rock and fluids can be expressed in terms of the contribution of each component's compressibility by volumetric phase saturation weighting as follows[1]:

$$c_t = S_o c_o + S_w c_w + S_g c_g + c_f \qquad \text{(E-1)}$$

where:

c_t = total system compressibility, psi^{-1}.
S_o = average oil saturation, fraction.
c_o = oil compressibility, psi^{-1}.
S_w = average water saturation, fraction.
c_w = water compressibility, psi^{-1}.
S_g = average gas saturation, fraction.
c_g = gas compressibility, psi^{-1}.
c_f = rock compressibility adjusted for porosity, psi^{-1}.

By definition the isothermal compressibility of a component c_n, is[1]:

$$c_n = -\frac{1}{V_n}\left(\frac{\delta V_n}{\delta p}\right)_T \qquad \text{(E-2)}$$

where:

V_n = volume of component n.

$(\delta V_n/\delta p)_T$ = rate of change in the volume of component n as a function of pressure at constant temperature.

As presented by Ramey,[1] the following expressions may be developed for the separate phase compressibilities:

$$c_o = -\frac{1}{V_o}\left(\frac{\delta V_o}{\delta p}\right)_T = -\frac{1}{B_o}\left(\frac{\delta B_o}{\delta p}\right)_T \qquad \text{(E-3)}$$

$$c_w = -\frac{1}{V_w}\left(\frac{\delta V_w}{\delta p}\right)_T = -\frac{1}{B_w}\left(\frac{\delta B_w}{\delta p}\right)_T \tag{E-4}$$

$$c_g = -\frac{1}{V_g}\left(\frac{\delta V_g}{\delta p}\right)_T = -\frac{1}{B_g}\left(\frac{\delta B_g}{\delta p}\right)_T \tag{E-5}$$

where:

B = formation volume factor.
V = volume of a particular fluid.

Inasmuch as gas is soluble in both oil and water, as well as being compressible, the phase compressibility terms for oil and water must also account for solution gas effects. The oil and water formation volume factors reflect the effect of solution gas on the change in liquid phase volumes. Thus, terms to reduce the gas phase volume by the quantity of gas going into solution, as pressure is raised, must be introduced for systems below the bubble point of the reservoir. This change in gas volume, ΔV_g, due to the solution of gas in either the oil or water phase may be represented as follows[1]:

$$\Delta V_g = S_n \frac{B_g}{B_n}\left(\frac{\delta R_n}{\delta p}\right)_T \tag{E-6}$$

where:

ΔV_g = change in gas volume, reservoir barrels/unit pore volume/psi.
S_n = saturation of phase n, res. bbl/res. bbl PV.
B_g = gas formation volume factor, res. bbl/SCF.
B_n = formation volume factor of phase n, res. bbl/STB.
$(\delta R_n/\delta p)_T$ = change of gas volume in solution with pressure in a particular phase, SCF/STB/psi.

Expression of this term for oil and water phases is as follows:

$$\Delta V_{go} = \frac{S_o B_g}{B_o}\left(\frac{\delta R_s}{\delta p}\right)_T \tag{E-7}$$

and

$$\Delta V_{gw} = \frac{S_w B_g}{B_w}\left(\frac{\delta R_{sw}}{\delta p}\right)_T \tag{E-8}$$

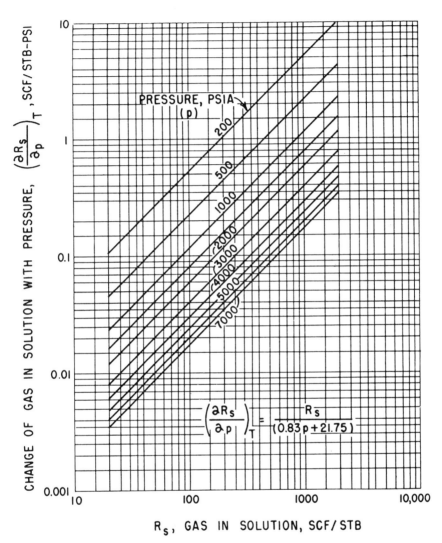

Fig. E-1. Relationship between solubility of gas in oil (SCF/STB) and change of gas in solution with pressure (SCF/STB-psi). (After Ramey,[1] courtesy of AIME.)

where:

R_s = solubility of gas in oil, SCF/STB.
R_{sw} = solubility of gas in water, SCF/STB.
B = formation volume factor for oil (o) or water (w), res. bbl/STB.
S = fluid saturation for oil (o) or water (w), fraction.

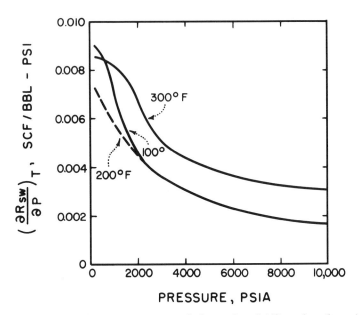

FIG. E-2. Relationship between pressure and change in solubility of methane in pure water. (After Ramey,[1] courtesy of AIME.)

The values of $(\delta R_s/\delta p)_T$ and $(\delta R_{sw}/\delta p)_T$ are preferably obtained from laboratory studies. If laboratory data are not available, the values of $(\delta R_s/\delta p)_T$ may be determined by use of Fig. E-1, which is based upon Standing's correlations for California crude oils.[2]

Inasmuch as the solubility of hydrocarbons in water decreases rapidly with increasing molecular weight, only the contributions from methane and ethane need to be considered. The expression for any mixture of ethane and methane is:

$$\left(\frac{\delta R_{sw}}{\delta p}\right)_T = y_1 \left(\frac{\delta R_{sw1}}{\delta p}\right)_T + y_2 \left(\frac{\delta R_{sw2}}{\delta p}\right)_T \tag{E-9}$$

where:

y_1 = mole fraction of methane.
$(\delta R_{sw1}/\delta p)_T$ = change of solubility of methane fraction, SCF/bbl-psi (see Fig. E-2).
y_2 = mole fraction of ethane.
$(\delta R_{sw2}/\delta p)_T$ = change of solubility of ethane fraction, SCF/bbl-psi (see Fig. E-3).

For temperatures below 250°F, one can use Fig. E-4; however, a correction to $(\delta R_{sw}/\delta p)_T$ for total solids in the brine should also be made (Fig. E-5).

Combining the relationships expressed by Eqs. E-1, E-3, E-4, E-5, E-7, and E-8, Martin's expression (extension of Perrine's expression) for total system compressibility, c_t, for a multiphase flow of water, oil and gas is[5]:

$$c_t = \frac{S_o}{B_o}\left[B_g\left(\frac{\delta R_s}{\delta p}\right) - \left(\frac{\delta B_o}{\delta p}\right)\right]_T + \frac{S_w}{B_w}\left[B_g\left(\frac{\delta R_{sw}}{\delta p}\right) - \left(\frac{\delta B_w}{\delta p}\right)\right]_T$$

$$- \frac{S_g}{B_g}\left(\frac{\delta B_g}{\delta p}\right)_T + c_f \qquad (E\text{-}10)$$

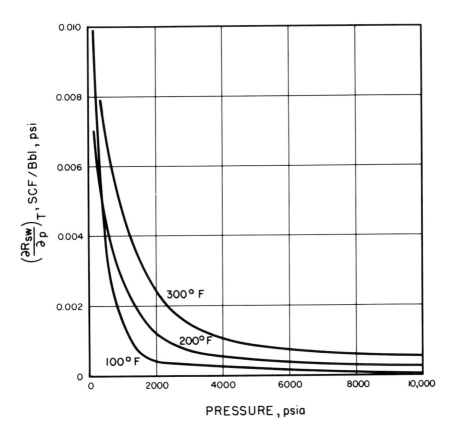

FIG. E-3. Relationship between pressure and change of solubility of ethane in pure water with pressure. (After Ramey,[1] courtesy of AIME.)

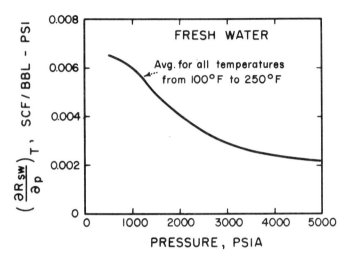

Fig. E-4. Relationship between pressure and change of natural gas in solution in fresh water, at 100–250°F temperature range for a natural gas containing 88.51% methane and 6.02% ethane.[3] Average error for 150–250°F is 9.5%, whereas the maximum error is ±15%. (After Ramey,[1] courtesy of AIME.)

Gas Reservoirs

The compressibility of natural gas ranges from 10^{-3} at 1,000 psi to about 10^{-4} at 5,000 psi. The sandstone and limestone compressibilities range from 3×10^{-6} to 10×10^{-6} (pore volume/pore volume) psi^{-1};

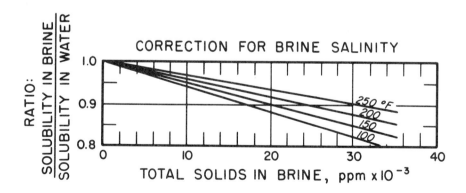

Fig. E-5. Correction factor to $(\delta R_{sw}/\delta p)_T$ for salinity of brines. (After Ramey,[1] courtsey of AIME.)

Rock and Fluid Compressibilities

whereas those of water with dissolved gas range from 15×10^{-6} psi^{-1} at 1,000 psi to 5×10^{-6} psi^{-1} at 5,000 psi.[1] Inasmuch as compressibilities for oil, water, and reservoir rock are small in comparison to that of the natural gas, the effective total system compressibility may be considered as being equal to:

$$c_t = S_g c_g \qquad (\text{E-11})$$

An approximate estimate may be made of the gas compressibility by considering it equal to the reciprocal of the absolute pressure.[1] This approximation is rigorously true for a perfect gas. It should not be used for natural gas at pressures above 500 psi.

$$c_t = \frac{1}{p_a} \qquad (\text{E-12})$$

where:

p_a = absolute pressure, psia.

Oil Reservoirs above the Bubble Point

Free gas saturations in reservoirs above the bubble point are generally small and although the values of compressibility are high for gas, its contribution to the total system compressibility is also relatively small. Consequently, the contributions of the rock, water and oil phases must be considered.

As long as the system remains above the bubble point, the quantity of solution gas dissolved in the fluids will remain approximately constant. The equation for the total system compressibility may be simplified from equation E-10 to read:

$$c_t = -\frac{S_o}{B_o}\left(\frac{\delta B_o}{\delta p}\right)_T - \frac{S_w}{B_w}\left(\frac{\delta B_w}{\delta p}\right)_T - \frac{S_g}{B_g}\left(\frac{\delta B_g}{\delta p}\right)_T + c_f \qquad (\text{E-13})$$

Oil Reservoirs below the Bubble Point

As the reservoir pressure declines from the initial pressure, p_i, to a pressure below the bubble point, p, the reservoir fluids and rock will expand. The difference between the system below the bubble point pressure and that above it is that below the bubble point pressure solution gas is liberated in the reservoir as pressure decreases. Adjustments for the oil and water compressibilities are required as illustrated by Eq. E-10. The adjusted total system compressibilities are often from 10 to 15 times greater than the unadjusted total system compressibilities, depending upon the reservoir temperature, pressure and the API gravity of the oil. The higher the oil gravity, the higher the solution gas content and the greater the impact upon compressibility. The amount of correction increases with the increase in the difference between the reservoir pressure and the bubble point pressure.

Composite equation E-10 should be used to evaluate the compressibility for oil, water and gas below the bubble point pressure.

Oil Reservoirs where Gas Saturation is Less than Critical

Ramey[1] has developed the following equation to find the total system compressibility when the gas saturation is less than the critical saturation for free gas flow:

$$c_t = -\frac{S_o}{B_o}\left(\frac{\delta B_t}{\delta p}\right) + \frac{S_w}{B_w}\left[B_g\left(\frac{\delta R_{sw}}{\delta p}\right)_T - \left(\frac{\delta B_w}{\delta p}\right)_T\right] + c_f \quad \text{(E-14)}$$

where:

B_t = total or two-phase (gas and oil) formation volume factor res. bbl/STB, or

$$B_t = B_o + (R_{si} - R_s)B_g \quad \text{(E-15)}$$

R_{si} = initial, or bubble-point solution gas/oil ratio, SCF/STB oil.
The value of $(\delta B_t/\delta p)_T$ may be obtained from Fig. E-6, which is based upon Standing's two-phase volume factors.

Calculation of Compressibilities for the Various Phases

Gas Phase Compressibility (c_g)

From the pseudo-critical values of temperature, T_c, and pressure, p_c, the gas deviation factor and its slope (dz/dp) may be calculated as follows:

(1) From Fig. D-3, p. 200, obtain the critical values of pressure, p_c, and temperature, T_c.

Rock and Fluid Compressibilities

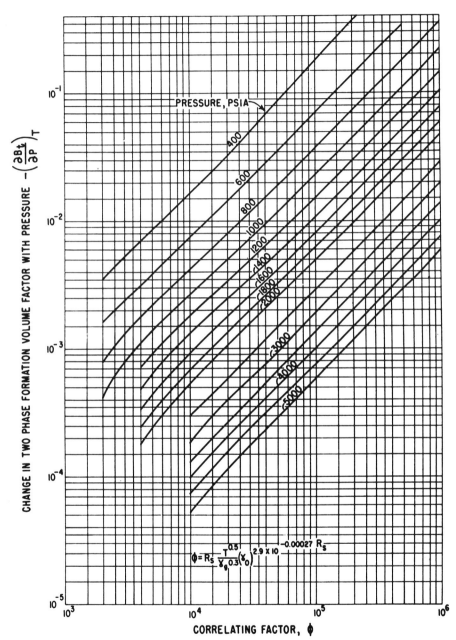

FIG. E-6. Change of Standing's[2] two-phase formation volume factors with pressure. (After Ramey,[1] courtesy of AIME.)

(2) Calculate the pseudo-reduced values of the temperature, T_r, and pressure, p_r:

$$T_r = \frac{T + 460}{T_c} \qquad \text{(E-16)}$$

$$p_r = \frac{p}{p_c} \qquad \text{(E-17)}$$

where:

T = reservoir temperature, °F.
p = reservoir pressure, psia.

(3) From Fig. C-1, p. 184, obtain the value of z and the slope dz/dp for the given pressure.

The value of c_g can be expressed in terms of the pressure and gas deviation factor, z, as follows[6]:

$$c_g = \frac{1}{p} - \frac{1}{z}\left(\frac{dz}{dp}\right) \qquad \text{(E-18)}$$

where:

dz/dp = the slope of the gas deviation factor versus pressure curve at pressure p.

Below pressures of 500 psi, Eq. E-18 may be simplified as follows[1]:

$$c_g \approx \frac{1}{p} \qquad \text{(E-19)}$$

The formation volume factor for gas may be computed from curves supplied in Appendix C, p. 183.

Oil Phase Compressibility (c_o)

The oil phase compressibility may be determined directly from laboratory data or use of Fig. E-7 which is a correlation presented at a course of advanced reservoir engineering at Texas A&M in 1964.

A third method of obtaining oil compressibility above the bubble point involves the use of the following equation[6]:

$$c_o = -\frac{(B_{o1} - B_{o2})}{B_{o1}(p_1 - p_2)} \qquad \text{(E-21)}$$

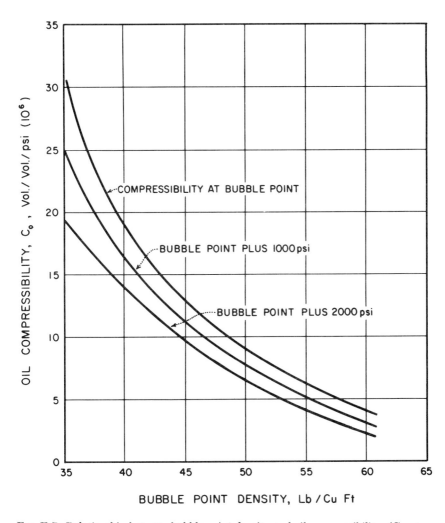

Fig. E-7. Relationship between bubble point density and oil compressibility. (Courtesy of M. B. Standing,[4] Standard Oil Co. of California.)

where the subscript "1" refers to time-1 and higher pressure, whereas "2" refers to time-2 at a lower pressure. The compressibility of undersaturated oils ranges from less than 5×10^{-6} to 30×10^{-6} psi^{-1}. Compressibility is greater for oils with a higher °API gravity and generally increases with increasing temperature. The oil formation volume factor may be calculated by techniques suggested in Appendix D.

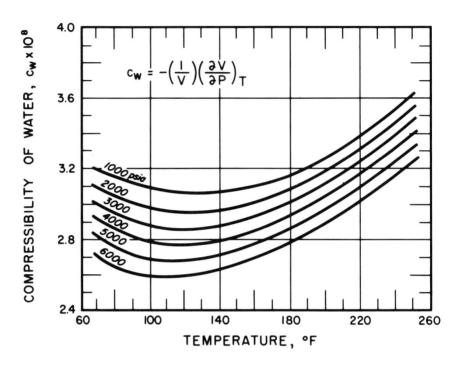
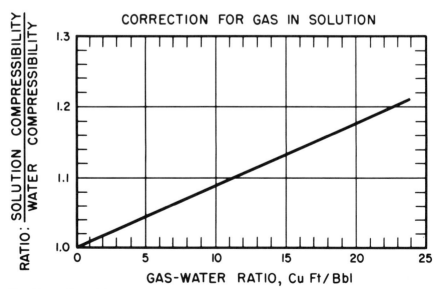

Fig. E-8. Effect of dissolved gas on compressibility of water. (After Dodson and Standing,[3] courtesy of API.)

Water Phase Compressibility (c_w)

Specific laboratory data for the compressibility of formation water, c_w, is often not available. Dodson and Standing[3] investigated compressibilities of water with varying salinity and natural gas content. The compressibilities of waters commonly encountered in the oil fields may be determined from Fig. E-8.

Rock Compressibility (c_f)

Hall[7] has presented a correlation between effective rock compressibility, c_f, and porosity (Fig. E-9). This graph is not applicable to either unconsolidated sands or highly fractured formations. In such cases, preferably laboratory data should be used.

Some data on compressibility of consolidated and unconsolidated rocks have been presented by Fatt,[8] van der Knaap and van der Vlis,[9] Sawabini,[10]

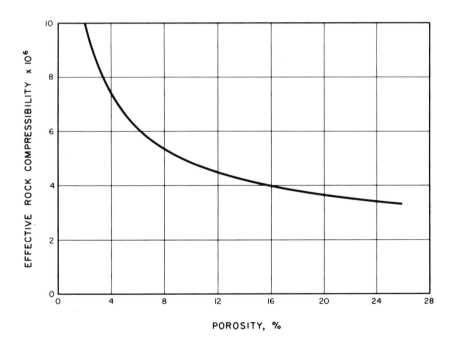

FIG. E-9. Relationship between porosity in percent and effective rock compressibility in psi^{-1} (change in pore volume/unit pore volume/psi). (After Hall,[7] courtesy of AIME.)

and others. The pore volume compressibilities

$$\left[c_b = -\frac{1}{V_b}\left(\frac{\delta V_b}{\delta p_e}\right)_{p_t,T}\right]$$

of unconsolidated sands determined by Sawabini[10] in a triaxial compaction apparatus ranged from 10^{-4} to 10^{-3} psi^{-1} in the 100 to 3,000 psi effective pressure range. The effective pressure was defined by him as a difference between the total overburden pressure, p_t, and the pore pressure p_p [$p_e = p_t - p_p$]. The tests were performed at constant overburden pressure of 3,000 psi and a temperature of 140°F. The bulk compressibilities

$$\left[c_b = -\frac{1}{V_b}\left(\frac{\delta V_b}{\delta p_e}\right)_{p_t,T}\right]$$

for the same pressure range varied from 7×10^{-4} to 3×10^{-5} psi^{-1}. It should be kept in mind that the compressibilities of sandstones obtained on using triaxial apparatus are about twice as high as those obtained in a uniaxial equipment.

Data on the compressibility of clays have been presented by Rieke et al.,[11,12] Chilingar et al.,[13,14] and Chilingarian and Rieke[15] (Tables E-1 and E-2).

TABLE E-1

Void Ratio and Compressibility Equations $\{c_b = -[1/(e + 1)](de/dp)\}$ for Various Clays Saturated in Water[a]

Type of clay	Assumed density (g/cc)	Relationship between void ratio and effective pressure	Compressibility equation (calculated)
Montmorillonite	2.60	$e = 2.69 - 0.467$ (log$p$)	$c_b = 3.25 \times 10^{-2}\, p^{-0.874}$ (above 1000 psi)
Illite	2.67	$e = 1.335 - 0.23$ (log$p$)	$c_b = 3.9 \times 10^{-2}\, p^{-0.926}$
Kaolinite	2.63	$e = 0.885 - 0.153$ (log$p$)	$c_b = 3.5 \times 10^{-2}\, p^{-0.946}$
Dickite	2.60	$e = 0.682 - 0.128$ (log$p$)	$c_b = 3.05 \times 10^{-2}\, p^{-0.980}$
Halloysite	2.55	$e = 1.01 - 0.165$ (log$p$)	$c_b = 3.3 \times 10^{-2}\, p^{-0.946}$
Hectorite (with 56% CaCO$_3$)	2.66	$e = 0.718 - 0.123$ (log$p$)	$c_b = 2.85 \times 10^{-2}\, p^{-0.954}$
P-95 dry lake clay (Buckhorn Lake, Calif.)	2.53	$e = 0.7 - 0.116$ (log$p$)	$c_b = 2.8 \times 10^{-2}\, p^{-0.946}$
Soil from limestone terrain (Louisville, Kentucky)	2.67	$e = 0.5 - 0.0816$ (log$p$)	$c_b = 2.25 \times 10^{-2}\, p^{-0.952}$

[a] After Chilingar et al.,[14] Chilingarian and Rieke,[15] and Rieke et al.;[11,12] recalculated by the writers.

TABLE E-2

Compressibility Equations [$c_b = -(1/h)(dh/dp)$] of Various Clays Calculated from Thickness (h) versus Overburden Pressure (p) Measurements[11,13]

Type of clay	Condition: dry or saturated	Compressibility equations obtained using different varieties of same clay
Dickite	Dry	$c_b = 3.96 \times 10^{-2} p^{-0.8925}$
Halloysite	Dry	$c_b = 2.27 \times 10^{-2} p^{-0.733}$
Hectorite (containing 60% $CaCO_3$)	Dry	$c_b = 4.15 \times 10^{-2} p^{-0.9825}$
Illite	Dry	$c_b = 3.4 \times 10^{-2} p^{-0.982}$
Kaolinite	Dry	$c_b = 3.58 \times 10^{-2} p^{-0.9091}$
Montmorillonite	Dry	$c_b = 3.29 \times 10^{-2} p^{-0.9348}$
Dickite	Distilled water	$c_b = 3.86 \times 10^{-2} p^{-0.924}; c_b = 3 \times 10^{-2} p^{-0.788}$
Halloysite	Distilled water	$c_b = 4.24 \times 10^{-2} p^{-0.848}; c_b = 4.7 \times 10^{-2} p^{-0.864}$
Hectorite (containing 62% $CaCO_3$)	Distilled water	$c_b = 3.38 \times 10^{-2} p^{-0.787}; c_b = 4.42 \times 10^{-2} p^{-0.830}$
Illite ⎫ (different varieties)	Distilled water	$c_b = 3.48 \times 10^{-2} p^{-0.815}; c_b = 3.73 \times 10^{-2} p^{-0.825}$
Illite ⎭	Distilled water	$c_b = 3.8 \times 10^{-2} p^{-0.93}$
Kaolinite	Distilled water	$c_b = 4.01 \times 10^{-2} p^{-0.811}; c_b = 3.48 \times 10^{-2} p^{-0.784}$
Montmorillonite[a]	Distilled water	$c_b = 2.78 \times 10^{-2} p^{-0.732}; c_b = 2.68 \times 10^{-2} p^{-0.735}$
Montmorillonite	Sea water	$c_b = 3.94 \times 10^{-2} p^{-0.902}$

[a] For the best straight line drawn through the experimental points ($p > 1000$ psi). Actually the curve is concave upwards.

Total System Compressibility (c_t)

The total system compressibility, c_t, commonly ranges from 10^{-5} to 10^{-4} for systems above the bubble point pressure.[1] When the system drops below the bubble point pressure, the compressibility increases as the pressure drops. The reason for this is the increase in the volume of the free gas phase with decreasing pressure below the bubble point. Other important references related to compressibilities, which should be consulted by the reader, include Geertsma,[16] van der Knaap,[17] Trube,[18] Long and Chierici,[19] and Allen.[20]

Compressibilities of Fractured-Cavernous Carbonates

An excellent discussion on compressibility of cavernous and fractured rocks is presented by Tkhostov et al.[21] According to them, at low effective pressures (up to 200–300 kg/cm²) the compressibility of fractures is of the order of 10^{-4} cm²/kg, whereas that of the caverns and vugs is of the order of 10^{-5} cm²/kg (cm²/kg = 7.031×10^{-2} psi^{-1}).

The compressibility of carbonate rock, c_{rc}, is equal to:

$$c_{rc} = (\phi_{fr} \times c_{fr}) + (\phi_{cv} \times c_{cv}) + c_m \qquad \text{(E-22)}$$

where:

ϕ_{fr} = fractional porosity of fractures.
c_{fr} = compressibility of fractures.
ϕ_{cv} = fractional porosity of caverns and vugs.
c_{cv} = compressibility of caverns and vugs.
c_m = compressibility of matrix.

For Solnhofen Limestone, $c_m = 0.03 \times 10^{-4}$ cm²/kg; whereas the compressibility of caverns and vugs, c_{cv}, was estimated by Tkhostov et al.[21] to be equal to:

$$c_{cv} \approx 3 \times c_m \approx 3 \times 0.03 \times 10^{-4} = 0.09 \times 10^{-4} \text{ cm}^2/\text{kg}$$

The following simplified equation for determining the secondary pore compressibility was derived by Tkhostov et al.[21]:

$$c_{ps} \leq \left[\left(\frac{\phi_{fr}}{\phi_{ts}} \times \frac{1350}{(p_t - p_p)} \right) - 0.09 \right] \times 10^{-4} \qquad \text{(E-23)}$$

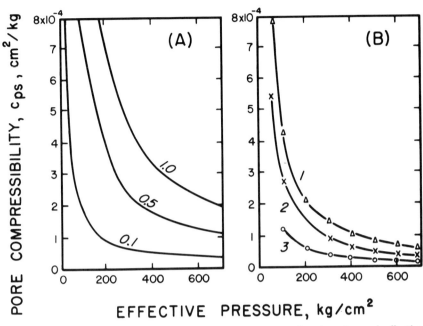

FIG. E-10. Relationship between the pore compressibility of carbonates and effective pressure ($p_e = p_t - p_p$). A—theoretical curves for fractured-cavernous reservoirs calculated on using Eq. E-23; numbers on curves designate the ϕ_{fr}/ϕ_{ts} ratios. B—experimental curves ($p_p = 0$) for (1) limestone with $\phi = 2.01\%$; (2) marl with $\phi = 2.63\%$; and (3) limestone with $\phi = 11.31\%$. (After Tkhostov et al.[21])

where:

c_{ps} = compressibility of secondary pores (fractures + vugs and caverns), cm²/kg.
ϕ_{ts} = total fractional porosity of secondary pores.
p_t = total overburden pressure, kg/cm².
p_p = pore pressure, kg/cm².

Relationship between the pore compressibility of carbonate rocks and effective overburden pressure is presented in Fig. E-10.

References

1. Ramey, H. J.: "Rapid Methods for Estimating Reservoir Compressibilities", *J. Pet. Tech.* (April, 1964) 447–454.
2. Standing, M. B.: *Volumetric and Phase Behavior of Oil Field Hydrocarbon Systems*, Reinhold Publ. Corp., New York (1952).

3. Dodson, C. R. and Standing, M. B.: "Pressure-Volume-Temperature and Solubility Relations for Natural Gas-Water Mixtures", *Drill. and Prod. Pract.*, API (1944) 173–179.
4. Standing, M. B., unpublished report (1964).
5. Martin, J. C.: "Simplified Equations of Flow in Gas Drive Reservoirs and the Theoretical Foundation of Multiphase Pressure Buildup Analysis", *Trans.*, AIME (1959) **216**, 309–311.
6. Craft, B. C. and Hawkins, M. F.: *Applied Petroleum Reservoir Engineering*, Prentice Hall, Inc., Englewood, N.J. (1959) 126, 149, 269.
7. Hall, H. N.: "Compressibility of Reservoir Rocks", *Trans.*, AIME (1953) **198**, 309–311.
8. Fatt, I.: "Pore Volume Compressibilities of Sandstone Reservoir Rocks", *Trans.*, AIME (1958) **213**, 362–364.
9. Van der Knaap, W. and van der Vlis, A. C.: "On the Cause of Subsidence in Oil-Producing Areas", *Seventh World Petrol. Cong.*, Mexico City, Elsevier Publ. Co. (1967) **3**, 85–95.
10. Sawabini, C. T.: *Triaxial Compaction of Unconsolidated Sandstone Core Samples Under Producing Conditions at a Constant Overburden Pressure of 3,000 psi and constant temperature of 140° F*, Ph.D. Dissertation, University of Southern California (1971).
11. Rieke, H. H. III, Ghose, S. K., Fahhad, S. A. and Chilingar, G. V.: "Some Data on Compressibility of Various Clays", *Proc. Internat. Clay Conf.* (1969) **1**, 817–828.
12. Rieke, H. H., III, Chilingar, G. V. and Robertson, J. O. Jr.: "High-Pressure (up to 500,000 psi) Compaction Studies on Various Clays", *Internat. Geol. Cong.*, 22nd Session, New Delhi, India (1964) Sec. 15, 22–38.
13. Chilingar, G. V., Rieke, H. H., III and Sawabini, C. T.: "Compressibilities of Clays and Some Means of Predicting and Preventing Subsidence", in: *Symposium on Land Subsidence*, Internat. Assoc. Sci. Hydrology and UNESCO, Tokyo, Japan (1969) **89**(II), 377–393.
14. Chilingar, G. V., Rieke, H. H., III and Robertson, J. O. Jr.: "Relationship Between High Overburden Pressures and Moisture Content of Halloysite and Dickite Clays", *Bull. Am. Geol. Soc.* (1963) **74**, 1041–1048.
15. Chilingarian, G. V. and Rieke, H. H., III, "Data on Consolidation of Fine-Grained Sediments", *J. Sediment. Petrol.* (1968) **33**, 811–816.
16. Geertsma, J.: "The Effect of Fluid Pressure Decline on Volumetric Changes of Porous Rocks", *Trans.*, AIME (1957) **210**, 331–340.
17. Van der Knaap, W.: "Nonlinear Behavior of Elastic Porous Media", *Trans.*, AIME (1959) **216**, 179–187.
18. Trube, A. S.: "Compressibility of Unsaturated Hydrocarbon Reservoir Fluids", *Trans.*, AIME (1957) **210**, 341–344.
19. Long, G. and Chierici, G.: "Salt Content Changes Compressibility of Reservoir Brines", *Petrol. Eng.* (July, 1961) B25–B31.
20. Allen, D. R.: "Physical Changes of Reservoir Properties Caused by Subsidence and Repressuring Operations", *J. Pet. Tech.* (Jan., 1968) 23–29.
21. Tkhostov, B. A., Vezirova, A. D., Vendel'shteyn, B. Yu. and Dobrynin, V. M.: *Oil in Fractured Reservoirs*, Izd. Nedra, Leningrad (1970) 173–197.

APPENDIX F

Relative Permeability Concepts

Introduction

The effective permeability of a porous medium to a fluid is a measure of the ability of the medium to transmit this fluid at the existing saturation, which is usually less than 100%. In the majority of formations there is a simultaneous existence of more than one phase: (1) oil and gas, (2) oil and water, or (3) oil, gas, and water. The concept of effective permeability implies that all but one phase are immobile. Inasmuch as a part of the effective pore space is occupied by another phase, a correction factor must be used. Permeability is measured by an arbitrary unit called "darcy", which is named after Henry d'Arcy. The permeability is equal to one darcy if 1 cm³ of fluid per second flows through 1 cm² of cross section of rock under a pressure gradient of 1 atm/cm, the fluid viscosity being 1 cp.

The magnitude of effective permeability depends on wettability, i.e., on whether (1) the immobile phase does not wet the solid surfaces of the rock and, therefore, occupies the central parts of the pores, or (2) the immobile phase wets the solid surfaces and thus tends to concentrate in smaller pores. The nature, distribution, and amount of immobile phase affects effective permeability.

The relative permeability to a fluid is defined as the ratio of effective permeability at a given saturation of that fluid to the absolute permeability at 100% saturation. The terms $k_{ro}(k_o/k)$, $k_{rg}(k_g/k)$, and $k_{rw}(k_w/k)$ denote the relative permeability to oil, to gas, and to water, respectively (k is the absolute permeability, often the single-phase liquid permeability). The relative permeabilities are expressed in percent or as a fraction.

In waterflooding projects or in natural water-drive pools, the relative permeabilities to oil and to water are of great importance. Where water and oil flow together, the relative permeabilities are affected by many factors which include (1) relative dispersion of one phase in the other, (2) time of contact with pore walls, (3) amount of polar substances in the oil, (4) degree of hardness of water, (5) relative amount of carbonate material in porous medium, and (6) temperature.

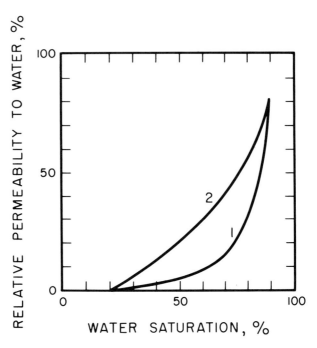

Fig. F-1. Relative permeability to water depending on speed of coalescence of water droplets and their attachment to solid surfaces. Curve 1, long time necessary for coalescence of water droplets together and their attachment to solid surfaces; and short time necessary for coalescence of oil droplets and their attachment to solid surfaces. Curve 2, short time necessary for coalescence and sticking of oil droplets to solid, surfaces and long time for water. (After Babalyan,[1] p. 141.)

The relative permeability to the continuous phase (dispersion medium) is greater than the relative permeability to the discontinuous (dispersed, internal) phase. With increasing degree of dispersion, the relative permeability increases for both continuous and discontinuous phases. The degree of dispersion increases with decreasing interfacial tension and increasing time of coalescence of dispersed-phase droplets.

Sticking (attachment) of the dispersed phase to solid surfaces depends on (1) interfacial tension, (2) angle of contact, (3) time necessary for the coalescence of droplets and lenses of the mobile part of a dispersed phase with an immobile part, and (4) thickness of dispersion medium (continuous phase) layer attached to solid surface. The relative permeability of the dispersed phase decreases if its droplets stick to the solid surface. The thickness of water film on solid surfaces is decreased in the presence of surface-active substances, which adsorb on the surfaces. In the case of alkaline water, which contains certain amounts of salts of organic acids (soaps), adsorbed layer (film) is thinner than in the case of hard or distilled water.

At low water saturations, water is present as a dispersed phase. The intensity of its transition from a dispersed phase into a dispersion medium (continuous phase) is determined by the intensity of coalescence of water droplets and of their sticking (attachment) to solid surfaces. The water saturation at which water changes from a dispersed phase into a continuous phase decreases with decreasing time of coalescence and sticking of water droplets to solid surfaces (Figs. F-1 and F-2).[1]

Effect of Polarity of Oil and Water Hardness on Relative Permeability Curves

With increasing concentration of polar substances in oil, the cumulative water production increases and the cumulative oil production decreases.[1]

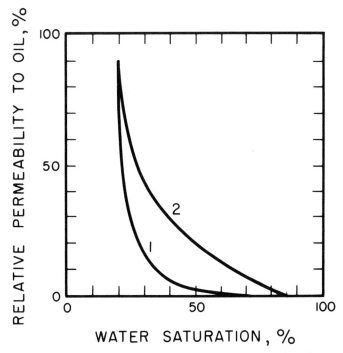

FIG. F-2. Relative permeability to oil as a function of speed of coalescence of oil droplets together and their attachment to solid surfaces. Curve 1, long time of coalescence of oil droplets and their attachment to solid surfaces; and short time necessary for coalescence of water droplets together and their attachment to solid surfaces. Curve 2, short time required for coalescence of oil droplets and their sticking to solid surfaces; and long, for water. (After Babalyan,[1] p. 142.)

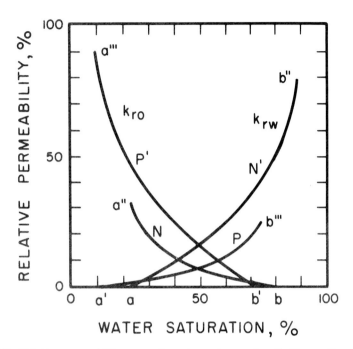

Fig. F-3. Relative permeability curves for polar and nonpolar oil. Curves P and P' are for polar oil, whereas curves N and N' are for nonpolar oil. (After Babalyan,[1] p. 145.)

The change in oil production rate upon increasing the concentration of polar substances in oil is quite rapid initially; then it slows down and eventually stabilizes when the polarity of oil reaches a certain limit. As shown in Fig. F-3, the critical saturation for water decreases (point a moves to a') and that for oil increases (point b moves to b') with increasing concentration of polar substances in oil. With decreasing concentration of polar substances in oil, the relative permeability to water sharply increases (point b''' moves to b''), whereas that to oil decreases (point a''' moves to a''). This is due mainly to the fact that attraction of nonpolar oil to solid surfaces is negligible and that the mobile oil presents less resistance to flow of water than does immobile oil.

According to Babalyan,[1] in the case of polar oil the water and oil production is greater when the water is alkaline than when it is hard.* This

---

* For alkaline water: $\dfrac{rNa^+}{rCl^-} > 1$; $\dfrac{rNa^+ - rCl^-}{rSO_4^{--}} > 1$; $rNa^+ + rCl^- > rSO_4^{--}$;

$rHCO_3^- > rCa^{++} + rMg^{++}$; $rNa^+ + rK^+ > rCl^- + rSO_4^{--}$.

is due to the change in the critical saturations of both phases (Fig. F-4). As shown in Fig. F-4, the relative permeability curves of oil + alkaline water lie above those of oil + hard water, because the following is true in the case of alkaline waters: (1) low interfacial tension between oil and water, (2) low values of contact angle, (3) slow coalescence of oil droplets in water, and (4) greater degree of dispersion of oil in water. In the case of hard water, on the other hand, the oil becomes a dispersed phase at higher water saturations in a porous medium than in the case of alkaline waters. The intensity of the transformation of oil into a dispersed phase is greater in alkaline than in hard waters.

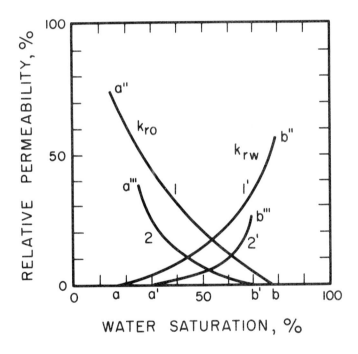

FIG. F-4. Relative permeabilities to oil and to water for polar oil + alkaline water (curves 1 and 1') and for polar oil + hard water (curves 2 and 2'). (After Babalyan,[1] p. 148.)

For hard water: $rNa^+/rCl^- < 1$; $\dfrac{rCl^- - rNa^+}{rMg^{++}} > 1$;

$$rCl^- + rSO_4^{--} > rNa^+ + rK^+.$$

(r = % equivalents)

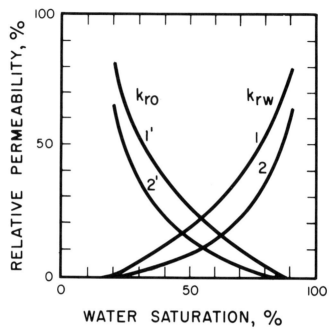

Fig. F-5. Relative permeability curves for nonpolar oil with alkaline water (curves 1 and 1′) and for nonpolar oil with hard water (curves 2 and 2′). (After Babalyan,[1] p. 149.)

In the case of nonpolar oil, the attachment of oil to solid surfaces is negligible in the presence of both alkaline and hard waters. When nonpolar oil flows with either alkaline or hard water, there is no change in critical saturations and hence the recovery of oil and water is the same in each case. The relative permeability curves with alkaline water, however, lie above those with hard water (Fig. F-5).

Effect of Water Chemistry on Recovery

Bernard[2] investigated the relative effectiveness of fresh and salt waters in flooding oil from cores containing clays. Experimental results indicated that when hydratable clays[3] are present, more oil is produced with fresh-water flood than with a brine flood.[2] The fresh-water flood, however, is accompanied by lowering in permeability and development of relatively high pressure drop. It is probable that the results obtained by Bernard[2] are owing to the effect of chemistry of water on relative permeabilities

Relative Permeability Concepts

(Fig. F-4), as previously discussed. Further research work should be done on the possibility of increasing oil recovery by adjusting chemistry of injection waters (relative proportions of Na^+, Cl^-, SO_4^{--}, HCO_3^-, Ca^{++}, and Mg^{++} ions; Na^+/Cl^-, $(Na^+-Cl^-)/SO_4^{--}$, Ca^{++}/Mg^{++}, Ca^{++}/Na^+, Ca^{++}/Cl^-, $(Cl^--Na^+)/Mg^{++}$ ratios; pH; etc.).

Effect of Carbonate Material in Porous Medium on Relative Permeabilities

The wettability of carbonates (calcite and dolomite) by oil is greater than that of quartz. Hence, the speed of adsorption of oil droplets to carbonate solid surfaces is greater than to quartz surfaces.[3] This hinders the movement of oil in a porous medium containing carbonate particles or cement. As the solid particles become more hydrophobic, the water saturation at which there is a transition of water from a dispersed phase

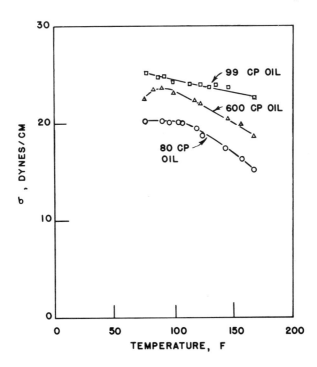

FIG. F-6. Relationship between interfacial tension and temperature. (After Poston et al.,[5] courtesy of *Soc. Pet. Eng. J.* of AIME.)

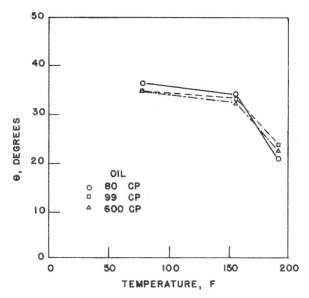

Fig. F-7. Relationship between contact angle and temperature for oil–water–glass system. (After Poston et al.,[5] courtesy of Soc. Pet. Eng. J. of AIME.)

to a dispersion medium increases, and the relative permeability to water remains low over a wide interval of water saturations. Solid surfaces also become more hydrophobic with increasing concentration of polar substances in the oil. In the case of nonpolar oil, the relative permeability to water is higher.

Effect of Temperature on Relative Permeability

The injection of hot fluids into an oil reservoir is becoming a very important oil recovery process. Poston et al.[5] and Sinnokrot et al.[6] have studied the effect of temperature on relative permeability. The following data are summarized from these studies. Experiments were carried out for clean quartz sand and natural oil sands, at temperatures ranging from room temperature to approximately 300°F. The increased temperatures for the oil–water–solid systems studied resulted in decreased interfacial forces (Fig. F-6), increased water-wetness (Fig. F-7), and important increases in both oil and water relative permeabilities (Figs. F-8 and F-9).

Figure F-8 indicates that, with respect to the wetting-phase (water) relative permeabilities, there was generally a cross-over of the curves in the low-water saturation region. This is caused by the increase in irreducible water saturation with increasing temperature. The relative permeability to water at higher water saturations increased with temperature. It is interesting to note, however, that relative permeabilities to water, k_{rw}, at 220° and 275°F were almost identical for water saturations from 30 to 60%.

The relative permeability of the nonwetting phase (oil) also increased with temperature (Fig. F-8). No cross-over occurred because the residual

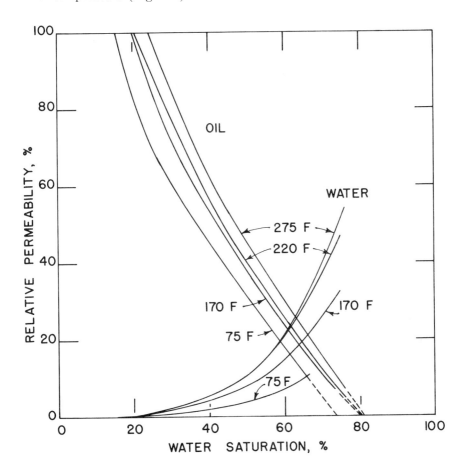

FIG. F-8. Relative permeability curves at different temperatures for Houston Sand and 80-cp oil. (After Poston et al.,[5] courtesy of *Soc. Pet. Eng. J.* of AIME.)

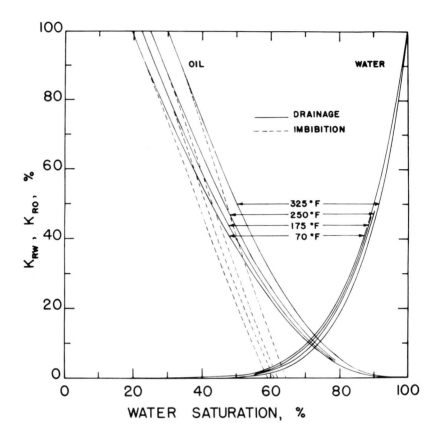

Fig. F-9. Calculated drainage and imbibition relative permeabilities for Bandera Sandstone core. (After Sinnokrot et al.,[6] courtesy of the SPE of AIME.)

oil saturation decreased with temperature increase. Poston et al.[5] noted that one interesting feature of the relative permeability curves is that the increase in both oil and water relative permeability at intermediate to high water saturations indicated a marked reduction in capillary forces within the porous media with temperature increase. The increase in total effective permeability, at 60% S_w, on raising temperatures from 75° to 275°F is about twofold under the influence of temperature alone. This, combined with a viscosity reduction of both oil and water by the temperature increase, would result in improved oil recoveries.

Relative Permeability Concepts

Figure F-10 shows relationship between the relative permeability ratio, k_{rw}/k_{ro}, and water saturation at various temperatures. Although a general increase in the relative permeability ratio occurs at temperatures above room temperature, there is no predictable trend.

The experiments showed that irreducible water saturation increases with increasing temperature when sandstone cores were used; this leads to the conclusion that capillary pressure–saturation curves are temperature sensitive (Figs. F-11, F-12, F-13, and F-14). Practically no change in S_{wi} with increasing temperature, however, was observed in the case of the

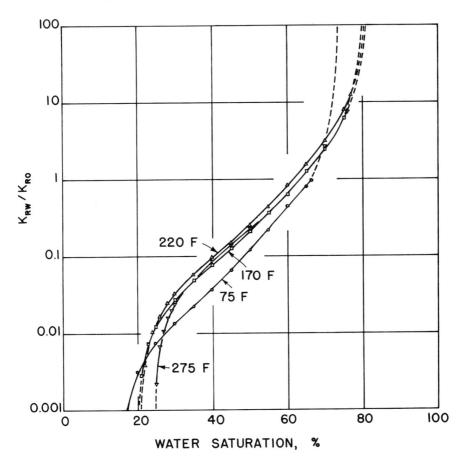

FIG. F-10. Relationship between k_{rw}/k_{ro} and water saturation at different temperatures for Houston Sand and 80-cp oil. (After Poston et al.,[5] courtesy of Soc. Pet. Eng. J. of AIME.)

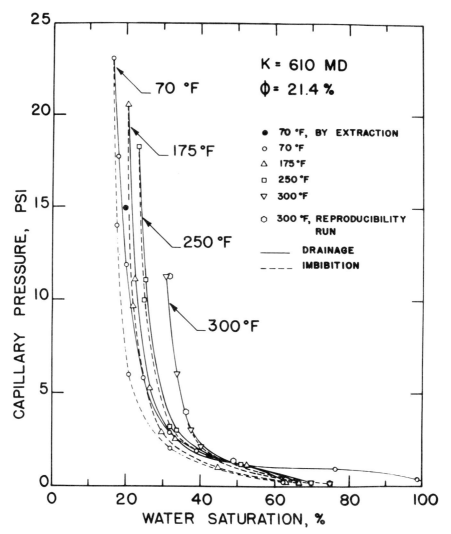

FIG. F-11. Capillary pressure curves for Berea Sandstone core at different temperatures. (After Sinnokrot et al.,[6] courtesy of the SPE of AIME.)

limestone core (Fig. F-14). These findings are of considerable importance in determination of irreducible water saturations used in estimation of recoverable oil. The "practical" oil saturation (residual oil at a WOR of 100) decreases with increasing temperature (Fig. F-15). The effect of temperature on the maximum water saturation on imbibition for sandstone and limestone cores is presented in Fig. F-16. All sandstone cores exhibited increases in the maximum imbibition water saturations at about the same rate as the increase in the irreducible water saturations, whereas the limestone core showed very little effect.

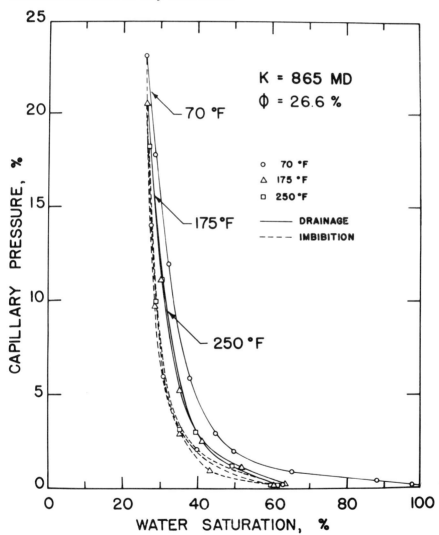

FIG. F-12. Capillary pressure curves for limestone core at different temperatures. (After Sinnokrot et al.,[6] courtesy of the SPE of AIME.)

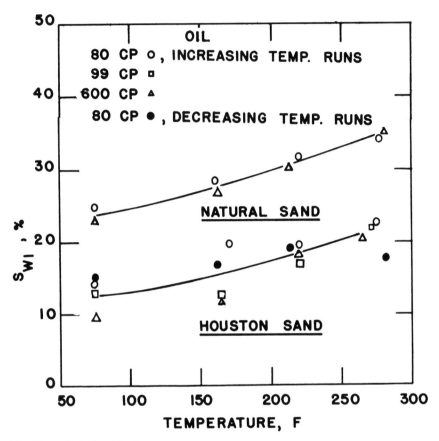

Fig. F-13. Relationship between irreducible water saturation and temperature for Houston sand and natural sand. (After Poston et al.,[5] courtesy of Soc. Pet. Eng. J. of AIME.)

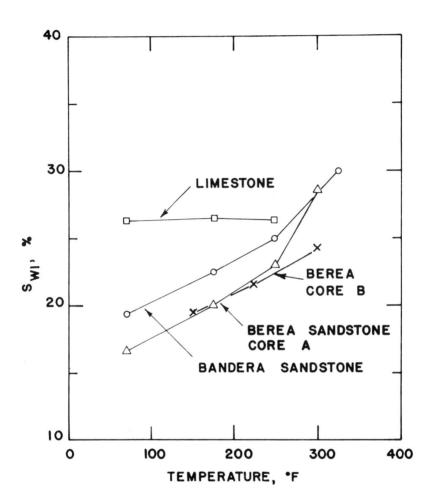

FIG. F-14. Relationship between irreducible water saturation and temperature for sandstone and limestone cores. (After Sinnokrot et al.,[6] courtesy of the SPE of AIME.)

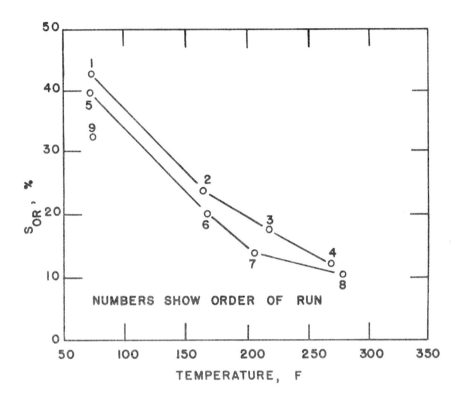

Fig. F-15. Relationship between residual oil saturation and temperature for Houston sand and 600-cp oil; effect of run sequence. Number indicates the order in which data were obtained. (After Poston et al.,[5] courtesy of *Soc. Pet. Eng. J.* of AIME.)

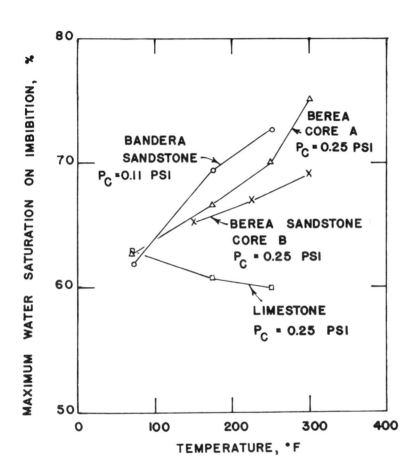

FIG. F-16. Relationship between maximum imbibition water saturation and temperature for sandstone and limestone cores. (After Sinnokrot et al.,[6] courtesy of the SPE of AIME.)

References

1. Babalyan, G. A.: *Questions on Mechanism of Oil Recovery*, Aznefteizdat, Baku (1956) 254 pp.
2. Bernard, G. C.: "Effect of Floodwater Salinity on Recovery of Oil from Cores Containing Clays", SPE paper 1725 presented at the AIME Meeting in Los Angeles, California (Oct., 1967) 8 pp.
3. Robertson, J. O. Jr., Rieke, H. H. III and Chilingar, G. V.: "Determination of Mineralogic Type of Clays from Degree of Hydration", *Sedimentology* (1965) **4,** 181–187.
4. Sinnokrot, A. and Chilingar, G. V.: "Effect of Polarity of Oil and Presence of Carbonate Particles on Relative Permeability of Rocks", *The Compass of Sigma Gamma Epsilon* (1961) **38,** 115–120.
5. Poston, S. W., Ysrael, S., Hossain, A. K. M. S., Montgomery, E. F. III and Ramey, H. J. Jr.: "The Effect of Temperature on Irreducible Water Saturation and Relative Permeability of Unconsolidated Sands", *Soc. Pet. Eng. J.* (June, 1970) **10,** No. 2, 171–180.
6. Sinnokrot, A. A., Ramey, H. J. Jr., and Marsden, S. S.: "Effect of Temperature Level Upon Capillary Pressure Curves", SPE paper 2517 presented at 44th SPE of AIME Annual Fall Meeting in Denver, Colorado (Sept. 28–Oct. 1, 1969).

APPENDIX G

Relationships Among Surface Area, Permeability, and Porosity

Introduction

Specific surface area, s_b, is defined as the surface of the pore channels per unit bulk volume, and is dependent upon the number, shape, size, and length of the pore channels. Specific surface area may be determined by laboratory analysis, a theoretical approach relating surface area to porosity and permeability, or a statistical method such as the one proposed by Chalkley et al.[1] A second method of expressing surface area is to relate it to the pore volume rather than to the bulk volume, in this case, the notation s_p is used.

Derivation of Theoretical Equation Relating Porosity, Permeability, and Surface Area

In a reservoir modeled by a bundle of capillary tubes, the rate of flow, q, is given by the Hagen–Poiseuille equation:

$$q = \frac{N\pi r^4 \Delta p}{8\mu L} \qquad (G\text{-}1)$$

where:

q = volumetric flow rate, cm³/sec;
N = number of capillaries;
r = capillary radius, cm;
Δp = differential pressure across the capillaries, dynes/cm²;
μ = fluid viscosity, poise;
L = length of the capillaries, cm.

The Darcy equation for rate of flow, q, is

$$q = \frac{kA\Delta p}{\mu L} \qquad (G\text{-}2)$$

where:

q = volumetric rate of flow, cm³/sec;
k = permeability, darcy;
A = total cross section area, cm²;
Δp = differential pressure, atm;
μ = fluid viscosity, cp;
L = length of the flow path, cm.

If, instead, viscosity is expressed in poises and differential pressure in dynes/cm², then:

$$q = \frac{9.869 \times 10^{-9} k \Delta p}{\mu L} \qquad (G\text{-}3)$$

The porosity, ϕ_c, of this bundle of capillary tubes may be expressed as the capillary volume, V_c, per unit of bulk volume, V_b:

$$\phi_c = \frac{V_c}{V_b} = \frac{N\pi r^2 L}{AL} = \frac{N\pi r^2}{A} \qquad (G\text{-}4)$$

Thus, the total cross-sectional area, A, of the bundle of tubes is

$$A = \frac{N\pi r^2}{\phi_c} \qquad (G\text{-}5)$$

The average capillary tube radius, r, may be found by combining Eqs. G-1, G-2, and G-5:

$$r = 2\left(\frac{2k}{\phi_c}\right)^{1/2} \qquad (G\text{-}6)$$

The surface area per unit of pore volume, s_p, is given by:

$$s_p = \frac{N 2\pi r L}{N \pi r^2 L} = \frac{2}{r} \qquad (G\text{-}7)$$

On substituting the value of the capillary tube radius from Eq. G-6 into Eq. G-7, the surface area, s_p, can be expressed as:

$$s_p = \left(\frac{\phi_c}{2k}\right)^{1/2} \qquad (G\text{-}8)$$

Solving Eq. G-8 for permeability yields

$$k = \frac{\phi_c}{2s_p^2} \tag{G-9}$$

Inasmuch as a porous rock is more complex than a bundle of capillary tubes, a constant, K_{cf}, is introduced. Thus, the equation for permeability becomes

$$k = \frac{\phi_c}{K_{cf} s_p^2} \tag{G-10}$$

Equation G-10 is the familiar Kozeny–Carman equation.[2] Carman[2] has noted that the constant, K_{cf}, is actually a complex combination of two variables, i.e., a shape factor for the pores, s_{hf}, and a tortuosity factor, τ:

$$K_{cf} = (s_{hf})(\tau) \tag{G-11}$$

Tortuosity is equal to the square of the ratio of the effective length, L_e, to the length parallel to the overall direction of flow of the pore channels, L:

$$\tau = \left(\frac{L_e}{L}\right)^2 \tag{G-12}$$

Thus, the Kozeny–Carman constant, K_{cf}, is a function of both the shape of each particular pore tube and the orientation of the pore tube relative to the overall direction of fluid flow.

Several theoretical relationships between tortuosity, τ, and porosity have been developed for simplified models, two of which are presented below. They are not, however, applicable to more complex media.

$$\tau = (F\phi)^2 \quad \text{(after Wyllie and Rose[3])} \tag{G-13}$$

$$\tau = F\phi \quad \text{(after Cornell and Katz[4])} \tag{G-14}$$

where F is the formation resistivity factor and is expressed as

$$F = \frac{R_o}{R_w} \tag{G-15}$$

In the above equation, R_o is equal to the electrical resistivity of a formation 100% saturated with formation water and R_w is equal to the formation water resistivity. The formation factor embodies the effect of grain size, grain shape, grain distribution, and grain packing. The effect of the porosity of the formation is thus included in the formation factor.

On measuring the values for L_e and L by ionic transit time theory[5] or by determining the tortuosity from a pore distribution concept, using a capillary pressure curve,[6] one can determine the value of tortuosity. Studies by Winsauer et al.[5] on natural cores yielded the following equation:

$$\tau = (F\phi)^{1.2} \tag{G-16}$$

This equation has been confirmed by the independent work of several other authors.[6,7]

Wyllie and Spangler[8] determined the pore shape factor, s_{hf}, in the case of unconsolidated sphere and bead packs. They indicated that the s_{hf} "constant" ranges from 2.13 to 3.32. The basic assumption of these studies is that the length of the collective pore tubes varies considerably, but the cross sections of this series of pore tubes are all of the same shape. For a medium of uniform grains and pores (such as sphere and bead packs) this is probably fairly accurate, but for a nonuniform medium it is not true. With the data collected by Wyllie and Spangler,[8] but using a value for tortuosity of $(F\phi)^{1.2}$, rather than $(F\phi)^2$, the calculations of s_{hf} were found not to give a constant value at all, but a variable which increases with increasing K_{cf}. Wyllie and Spangler[8] calculated the values for K_{cf} after measuring the porosity, permeability, and surface area (by pindrop technique) of the natural specimens. Utilizing Eq. G-16, and using data by Wyllie and Spangler[8] and Wyllie and Gregory,[9] one can obtain a straight line on log–log paper having an equation of[10]

$$s_{hf} = 1.55 K_{cf}^{0.455} \tag{G-17}$$

Because s_{hf} and K_{cf} are interrelated by the tortuosity, τ, the expression for shape factor and the overall Kozeny–Carman factor can be expressed in terms of easily obtained parameters[10]:

$$s_{hf} = 2.24 F\phi \tag{G-18}$$

and

$$K_{cf} = 2.24 (F\phi)^{2.2} \tag{G-19}$$

The final expression for the surface area per unit of pore volume in cgs units (cm^2/cm^3) is, therefore:

$$s_p = \frac{2.11 \times 10^5}{(F^{2.2}\phi^{1.2}k)^{1/2}} \tag{G-20}$$

where the permeability, k, is expressed in millidarcys.[10]

Kotyakhov developed the following formula for surface area per unit of bulk volume (cm²/cm³) of sands[11,12]:

$$s_b = 7000 \left(\frac{\phi^3}{k}\right)^{1/2} \quad \text{(G-21)}$$

where ϕ is the fractional porosity and k is the permeability expressed in darcys. This equation is another version of the Kozeny–Carman equation:

$$k = \frac{10^8 \phi^3}{(s_{hf})(\tau)(s_b)^2} \quad \text{(G-22)}$$

For a consolidated rock ($s_{hf} = 2$) and assuming that $\tau = 1$, which, in effect, means that rock sample is equivalent to a bundle of capillary tubes, one obtains

$$k = \frac{10^8 \phi^3}{(2)(1)(s_b)^2} \quad \text{or} \quad s_b \approx 7000 \left(\frac{\phi^3}{k}\right)^{1/2}$$

It is clear that Kotyakhov's[12] equation is an oversimplification for a rock sample. If values of $\tau = 1.25$ and $s_{hf} = 2.5$, representative of unconsolidated sands or calcarenites, are inserted in Eq. G-22 the surface area may be expressed as

$$s_b = 5650 \left(\frac{\phi^3}{k}\right)^{1/2} \quad \text{(G-23)}$$

where:

s_b = surface area in cm² per unit (cm³) of bulk volume,
ϕ = fractional porosity,
k = permeability in darcys.

Equation G-23 should give a good approximation of surface area for unconsolidated sediments, but should not be used for consolidated sediments.

Pirson[13,14] proposed the following formula for determining surface area of carbonate rocks:

$$k = \frac{10^8}{2F s_p^2} \quad \text{(G-24)}$$

or

$$s_p = 10^4 \left(\frac{1}{2Fk}\right)^{1/2} \quad \text{(G-25)}$$

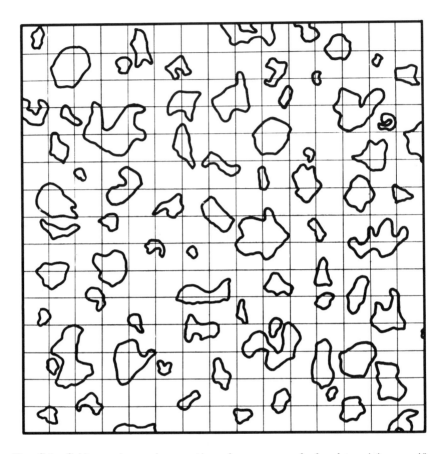

Fig. G-1. Grid superimposed on section of porous sample for determining specific surface area. (After Perez-Rosales,[16] courtesy of the SPE of AIME.)

where:

k = Klinkenberg permeability, d;
s_p = surface area per unit of pore volume, cm^2/cm^3;
F = formation resistivity factor (dimensionless) which is related to the porosity by the equation

$$F = \phi^{-m} \qquad (G\text{-}26)$$

where the cementation factor m varies from 1.3 for unconsolidated sands and oolitic limestones to 2.2 for dense limestones.

Shirkovskiy[15] proposed the following formula for determining specific area (surface area/unit bulk volume), s_b, of fragmental rock without interstitial water in cm²/cm³:

$$s_b = \frac{\phi^{3/2}}{\sqrt{2}k^{1/2}\tau^{1/2}} \tag{G-27}$$

where ϕ = porosity, fractional; k = permeability, perm.*; and τ = tortuosity. The specific surface area in cm²/cm³ with correction for interstitial water is equal to

$$s_{bw} = s_b(1 - S_w)^{3/4} \tag{G-28}$$

where S_w = interstitial water saturation, fraction.

Statistical Technique of Determining Specific Surface Area

A simple statistical method may be used as an independent check of values obtained for the surface area by laboratory and theoretical analyses. This technique, based on the work of Chalkley et al.,[1] consists of superimposing a grid system upon an enlarged reproduction of a thin section of the core as shown in Fig. G-1. A count is then made of the number of

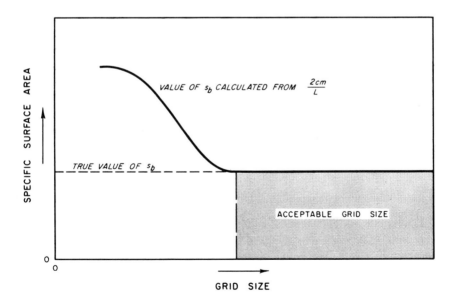

FIG. G-2. Determination of optimum grid size. (Modified after Perez-Rosales,[16] courtesy of the SPE of AIME.)

---

* Permeability in cgs units; 1 darcy = 1.02×10^{-8} perm.

TABLE G–1

Classification of Fragmental Carbonate Reservoir Rocks (Calcarenites, etc.)[a]

Class	Name of rock and granulometric composition	Effective porosity, %	Absolute permeability to gas, md	Characteristic of rock on the basis of porosity and permeability
I	Calcarenite, medium-grained (0.25–0.50 mm)	≥17	≥1000	Very good
	Calcarenite, fine-grained (0.1–0.25 mm)	≥20		
	Calcisiltite, coarse-grained (0.05–0.1 mm)	≥24		
	Calcisiltite, fine-grained (0.01–0.05 mm)	≥29		
II	Calcarenite, medium-grained	15–17	500–1000	Good
	Calcarenite, fine-grained	18–20		
	Calcisiltite, coarse-grained	22–24		
	Calcisiltite, fine-grained	27–29		
III	Calcarenite, medium-grained	11–15	100–500	Medium
	Calcarenite, fine-grained	14–18		
	Calcisiltite, coarse-grained	17–22		
	Calcisiltite, fine-grained	21–27		
IV	Calcarenite, medium-grained	6–11	10–100	Medium low
	Calcarenite, fine-grained	8–14		
	Calcisiltite, coarse-grained	10–17		
	Calcisiltite, fine-grained	12–21		

Relationships Among Surface Area, Permeability, and Porosity

TABLE G-1 (cont.)

Classification of Fragmental Carbonate Reservoir Rocks (Calcarenites, etc.)[a]

Class	Name of rock and granulometric composition	Effective porosity, %	Absolute permeability to gas, md	Characteristic of rock on the basis of porosity and permeability
V	Calcarenite, medium-grained	0.5–6	1–10	Poor
	Calcarenite, fine-grained	2–8		
	Calcisiltite, coarse-grained	3–10		
	Calcisiltite, fine-grained	4–12		
VI	Calcarenite, medium-grained	<0.5	<1	Very poor
	Calcarenite, fine-grained	<2		
	Calcisiltite, coarse-grained	<3.3		
	Calcisiltite, fine-grained	<3.6		

[a] Modified after Khanin,[17] p. 234.

intersections, c, of the grid lines with the perimeter of the pores. If the total length, L, of all the grid lines and the linear magnification, m, of the reproduction are known, the specific surface area, s_b, is[16]

$$s_b = \frac{2cm}{L} \tag{G-29}$$

To determine the optimum mesh or spacing between the grid lines, utilize several grid line mesh templates and make several calculations of the specific surface area. Plot the values of specific area, s_b, versus the number of grid intersections, c, or versus the grid size and select a grid which falls within the acceptable region as shown in Fig. G-2.

Relationships among Rock Granulometric Composition, Porosity, and Permeability

The relationship among rock granulometric composition, porosity, and permeability, modified after Khanin[17] is presented in Table G-1.

References

1. Chalkley, H. W., Cornfield, J. and Park, H.: "A Method for Estimating Volume-Surface Ratios", *Science* (Sept., 1949) **110,** 295.
2. Carman, P. C.: "The Determination of the Specific Surface of Powders", *Trans. Inst. Chem. Eng.*, London (1937) **15,** 150.
3. Wyllie, M. R. J. and Rose, W. D.: "Some Theoretical Considerations Related to the Quantitative Evaluation of the Physical Characteristics of Reservoir Rock from Electrical Log Data", *Trans.*, AIME (1950) **189,** 105–118.
4. Cornell, D. and Katz, D. L.: "Flow of Gases through Consolidated Porous Media", *Ind. and Eng. Chem.* (1953) **45,** 2145–2152.
5. Winsauer, W. O., Shearin, H. M., Jr., Masson, P. H. and Williams, M.: "Resistivity of Brine-Saturated Sands in Relation to Pore Geometry", *Am. Assoc. Pet. Geol. Bull.* (1952) **36,** 253–277.
6. Faris, S. R., Gournay, L B., Lipson, L. B. and Webb, T. S.: "Verification of Tortuosity Equations", *Am. Assoc. Pet. Geol. Bull.* (1954) **38,** 2226–2232.
7. Purcell, W. R.: "Capillary Pressures—Their Measurement Using Mercury and the Calculation of Permeability Therefrom", *Trans.*, AIME (1949) **186,** 39–48.
8. Wyllie, M. R. J. and Spangler, M. B.: "Application of Electrical Resistivity Measurements to Problem of Fluid Flow in Porous Media", *Am. Assoc. Pet. Geol. Bull.* (1952) **36,** 359–403.
9. Wyllie, M. R. J. and Gregory, A. R.: "Effect of Porosity and Particle Shape on Kozeny–Carman Constants", *Ind. and Eng. Chem.* (1955) **47,** 1379–1388.
10. Chilingar, G. V., Main, R. and Sinnokrot, A.: "Relationship between Porosity, Permeability, and Surface Areas of Sediments", *J. Sed. Petrol.* (1963) **33,** No. 3, 759–765.
11. Eremenko, N. A.: *Geology of Petroleum (Handbook)*, Tom I, GosTopTekhIzdat, Moscow (1960) 592 pp.
12. Kotyakhov, F. I.: "Relationship between Major Physical Parameters of Sandstones", *Neft. Khoz.* (1949) No. 12, 29–32.
13. Pirson, S. J.: *Oil Reservoir Engineering*, 2nd ed., McGraw-Hill, New York (1958) 101, 108.
14. Craft, B. C., Holden, W. R. and Graves, E. D., Jr.: *Well Design (Drilling and Production)*, Prentice-Hall, Englewood Cliffs, New Jersey (1962) 539.
15. Shirkovskiy, A. I.: "Complex Field Investigations in Gas Condensate Deposits of Cambay, India", *Geology, Exploration and Development of Gas and Gas-Condensate Deposits*, Vniie—Gazprom, Moscow (1969) 69, 70.
16. Perez-Rosales, C.: "A Simplified Method for Determining Specific Surface", *J. Pet. Tech.* (Aug., 1967) 1081–1084.
17. Khanin, A. A.: *Oil and Gas Reservoir Rocks and Their Study*, Izd. Nedra, Moscow (1969) 366 pp.

Bibliography

1. Greer, F. C. and Beeson, C. M.: "Permeability as a Function of Internal Surface Areas of Cores", Paper presented at 31st Annual California Meeting of SPE of AIME (1960) 7 pp.
2. Greer, F. C., Main, R., Beeson, C. M. and Chilingar, G. V.: "Determination of Surface Areas of Sediments", *J. Sed. Petrol.* (1962) **32,** 140–145.
3. Rose, W. and Bruce, W. A.: "Evaluation of Capillary Character in Petroleum Reservoir Rock", *Trans.*, AIME (1949) **186,** 127–142.

APPENDIX H

Carbonate Reservoir Data for Text Problems

A carbonate reservoir, A, with an intercrystalline–intergranular porosity system, was discovered in 1944. It was found at a vertical depth of 7300 to 8100 ft. Tables H-1 through H-7 and Figs. H-1 and H-2 summarize the reservoir rock and fluid properties. The reservoir data are modified after Welsh, J. R., Simpson, R. E., Smith, J. W., and Yust, C. S.: "A Study of Oil and Gas Conservation in the Pickton Field," *Trans.*, AIME (1949) 186, 55–65.

In solving the problems in the text, assume that nine of the wells are to be converted to injectors. Use a water injection rate of 1000 B/D per well or return 85% of the produced gas to the reservoir in the case of gas injection. Assume that injection project was started on November 1, 1947.

The k_g/k_o data can be selected by the instructor. The necessary data may also be obtained from references 2, 3, and 4.

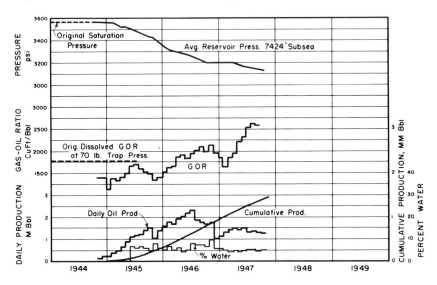

FIG. H-1. Monthly average reservoir production data. (After Welsh *et al.*,[1] courtesy of the SPE of AIME.)

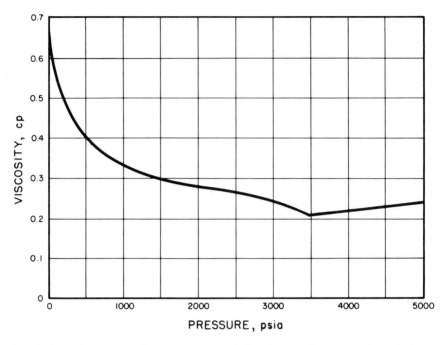

FIG. H-2. Relationship between pressure and viscosity of oil at reservoir conditions. (After Welsh et al.,[1] courtesy of the SPE of AIME.)

Carbonate Reservoir Data for Text Problems

TABLE H-1

General Reservoir Data for Carbonate Reservoir A[a]

Original oil–water contact, ft SS (subsea)	7,425
Average porosity, percent	19
Permeability, millidarcys[b]	
Parallel to bedding plane	379
Perpendicular to bedding plane	183
Connate water, percent of pore space	25
Average net pay section, ft	14
Original reservoir pressure, psig at 7424 ft SS	3,578
Reservoir temperature, °F	209
Characteristics of produced oil[c]	
Dissolved gas–oil ratio	2,220
Shrinkage factor, bbl stock-tank oil per reservoir bbl	0.413
Oil gravity, API	46.0
Sodium chloride content of connate water, ppm	260,000
Viscosity of original reservoir oil, cp	0.21

[a] After Welsh et al.,[1] courtesy of the SPE of AIME.
[b] Average of approximately 380 determinations on cores from 19 wells.
[c] Original reservoir oil produced under separator conditions of 0 psig and 80 °F.

TABLE H-2

Production and Reserve Data for Carbonate Reservoir A[a]

Estimated productive area, acres	2,600
Producing oil wells as of 11-1-47	36
Estimated oil-in-place, reservoir bbl	40,785,586
Cumulative oil production as of 11-1-47	1,382,694
October 1947 monthly oil production, bbl	42,481
Cumulative gas production as of 11-1-47, Mcf	2,578,631
Cumulative water production as of 11-1-47, bbl	81,743
Assumed abandonment reservoir pressure, psig	100

[a] After Welsh et al.,[1] courtesy of the SPE of AIME.

TABLE H–3

Pressure-Volume-Temperature Relations of Subsurface Saturated Oil Sample[a]

Temperature: 76 °F			Temperature: 150 °F		
Pressure, psig	Relative volume	Specific volume, cu ft/lb	Pressure, psig	Relative volume	Specific volume, cu ft/lb
4,500	.9681		4,500	.9679	
4,000	.9766		4,000	.9799	
3,800	.9804		3,800	.9850	
3,600	.9842		3,600	.9905	
3,428	.9875		3,408	.9962	
3,220	.9918		3,295	1.0000	0.02768
3,022	.9960		3,273	1.0022	
2,872	1.0000	0.02582	3,240	1.0057	
2,816	1.0050		3,200	1.0102	
2,769	1.0097		3,160	1.0146	
2,723	1.0143		3,050	1.0279	
2,643	1.0238		2,790	1.0680	
2,540	1.0380		2,510	1.1262	
2,230	1.0953		2,202	1.2161	
1,938	1.1814		1,908	1.3423	
1,630	1.3254		1,615	1.5282	
1,290	1.6045		1,315	1.8268	
1,050	1.9339		1,110	2.1474	
890	2.2618		963	2.4555	
780	2.5927		850	2.7755	
692	2.9208		770	3.0838	
568	3.5905		645	3.7143	
480	4.2601		550	4.3473	

[a] After Welsh et al.,[1] courtesy of the SPE of AIME.

TABLE H-3 (Continued)

Temperature: 179 °F			Temperature: 209 °F		
Pressure, psig	Relative volume	Specific volume, cu ft/lb	Pressure, psig	Relative volume	Specific volume, cu ft/lb
4,500	.9668		4,500	.9635	
4,000	.9804		4,000	.9791	
3,800	.9866		3,800	.9863	
3.600	.9933		3,610	.9938	
3,412	1.0000	0.02843	3,500	1.0000	0.02925
3,385	1.0027		3,478	1.0021	
3,345	1.0071		3,440	1.0063	
3,310	1.0114		3,407	1.0105	
3,272	1.0158		3,375	1.0148	
3,170	1.0288		3,200	1.0378	
2,895	1.0723		2,900	1.0898	
2,600	1.1335		2,600	1.1596	
2,300	1.2214		2,317	1.2467	
2,000	1.3447		2,008	1.3894	
1,622	1.5829		1,723	1.5717	
1,322	1.8942		1,535	1.7327	
1,130	2.1950		1,310	2.0020	
985	2.5066		1,122	2.3125	
878	2.8077		1,000	2.5918	
720	3.4190		908	2.8507	
615	4.0359		750	3.4428	
			648	4.0349	

TABLE H-4

Hydrocarbon Analysis of Subsurface Oil and Gas Sample[a]

Component	Original saturated material			Solution gas flashed at 0 psig & 140 °F		
	Wt. %	Density, g/cc at 60 °F	Molecular weight	Vol. %	Density, g/cc at 60 °F	Molecular weight
Methane	13.80			59.97		
Ethane	3.52			7.88		
Propane	5.88			9.97		
Isobutane	2.54			2.89		
n-Butane	5.75			6.77		
Isopentane	1.26			1.26		
n-Pentane	4.32			4.38		
Hexane	6.23			3.71		
Heptane	2.80	.7168	98			
Octane	5.42	.7346	105			
Nonane	6.61	.7611	111			
Heavier fraction	41.87	.8420	200	3.17	.7265	100
Total	100.00			100.00		
Hexane-free fraction (hexane and lighter)		.8133	163			

[a] After Welsh et al.,[1] courtesy of the SPE of AIME.

TABLE H-4 (Continued)

A.S.T.M. Distillation of Residual Oil Sample:*

% Over	Temp.: °F
I.B.P. **	110
5	160
10	186
15	220
20	251
25	276
30	297
35	333
40	362
45	398
50	432
55	468
60	508
65	550
70	595
75	645
80	685

Maximum temp: 686 °F
Recovery: 83 % by vol.
Residue: 15 % by vol.
Loss: 2 % by vol.
Gravity: API at 60 °F
   Distillation charge   48.0
   Overhead product   51.2

* Sample obtained by flash gas liberation, from subsurface oil sample having saturation pressure of 3,500 psig at 209 °F, to atmospheric pressure and 20 psig at 80 °F.
** Initial boiling point.

TABLE H-5

Summary of Volumetric Balance Calculations—Carbonate Reservoir A
(After Welsh et al.,[1] courtesy of SPE of AIME.)

Date	Reservoir pressure, psig @ 7,424' SS	Cum. oil, STB	Cum. oil, orig. res. bbl	Recovery, orig. oil-in-place	Cum. water prod., bbl
1944					
November 6	3,578	0	0	0	0
1945					
January 15	3,565	10,547	22,600	0.055	0
February 18	3,562	20,256	43,400	0.105	0
April 11	3,527	46,135	98,800	0.239	0
May 1	3,520	59,299	126,900	0.307	0
June 6	3,505	92,488	198,000	0.478	666
August 6	3,478	162,447	348,000	0.842	5,299
September 16	3,448	214,395	459,000	1.110	8,609
October 26	3,436	278,927	597,000	1.440	11,772
November 24	3,432	304,879	653,000	1.580	14,214
1946					
March 15	3,359	477,860	1,022,000	2.480	25,925
July 1	3,260	691,492	1,480,000	3.580	37,695
November 1	3,207	912,162	1,952,000	4.720	55,632
1947					
February 15	3,215	1,015,000	2,172,000	5.480	64,283
April 1	3,202	1,083,707	2,320,000	5.600	67,063
June 23	3,155	1,190,415	2,550,000	6.170	72,021

TABLE H-5 (Continued)

Cum. gas prod., Mcf	Avg. GOR between surveys, ft³/STB	Cum. total fluid withdrawals, bbl	Total water influx, bbl	Net water influx, bbl	bbl net water influx per bbl orig res. oil prod.
0	—	0	0	0	0.05
—	—	22,550	−31,150	−31,150	−1.38
—	—	43,400	−18,600	−18,600	−0.43
56,830	—	99,100	−115,900	−115,900	−1.17
76,063	1,366	127,600	−120,400	−120,400	−0.95
127,643	1,472	199,800	−118,200	−118,900	−0.60
243,766	1,658	356,300	−90,700	−96,000	−0.28
330,410	1,548	473,600	−121,400	−130,000	−0.28
412,693	1,523	617,800	−36,200	−48,000	−0.08
457,063	1,398	678,200	4,200	−21,700	−0.03
716,925	1,514	1,076,000	21,000	−16,700	−0.02
1,111,183	1,805	1,575,000	−50,000	−115,600	−0.08
1,551,858	1,970	2,100,000	165,000	109,400	+0.06
1,760,653	1,887	2,344,000	454,000	389,700	+0.18
1,873,591	1,887	2,504,500	537,500	470,400	+0.20
2,107,579	2,190	2,813,400	551,400	479,400	+0.19

TABLE H-6

Properties of Subsurface Oil Sample—Carbonate Reservoir A
(After Welsh et al.,[1] courtesy of SPE of AIME.)

Date taken: November 16, 1944
Sampling conditions: depth = 7,700 ft; well was shut in.
Properties of saturated oil:

Temperature, °F	76	150	179	209
Saturation pressure, psig	2872	3295	3412	3500

Gas liberation and shrinkage of oil

Flash vaporization

Pressure, psig (p_1)	Temp., °F	Gas/oil ratio, cu ft per bbl residual oil		Residual oil gr, °API at 60 °F	Sp gr gas at 60 °F	V_r/V_s*
		Flashed at p_1	Flashed from p_1 to 0			
0	80	2,220	0	46.0	1.018	.413
20	80	1,924	—	48.1		.453
150	80	1,497	241	50.5		.478
500	80	1,316	527	49.0		.464

Differential vaporization at 209 °F

Pressure, psig	Gas/oil ratio: cu ft at 60 °F. per bbl reservoir oil at 3,500 psig, 209 °F	Residual oil gravity, °API at 60 °F	V/V_s**
3,500	0		1.000
3,200	154		.9064
2,900	264		.8398
2,600	350		.7907
2,300	423		.7519
2,000	486		.7167
1,700	542		.6875
1,400	593		.6614
1,100	643		.6370
800	691		.6123
500	739		.5878
270	783		.5600
148	806		.5418
0	882	47.6	.4820

* V_r, volume residual oil at 0 psig, 60 °F.
 V_s, volume saturated oil at 3,500 psig, 209 °F.
** V, volume saturated oil at indicated pressure, 209 °F.

TABLE H-7

Permeability Data from Core Tests for Carbonate Reservoir A

Permability, md	No. of samples [a]
730	1
598	1
458	2
450	1
400	1
375	3
325	1
300	2
255	1
207	1

[a] Assume that each sample represents 1-ft interval of the net pay section.

References

1. Welsh, J. R., Simpson, R. E., Smith, J. W. and Yust, C. S.: "A Study of Oil and Gas Conservation in the Pickton Field", *Trans.*, AIME (1949) **186**, 55–65.
2. Chilingar, G. V., Mannon, R. W. and Rieke III, H.: Oil and Gas Production from Carbonate Rocks, American Elsevier Publ. Co., New York (1971).
3. Stewart, C. R., Craig, F. F. and Morse, R. A.: "Determination of Limestone Performance Characteristics by Model Flow Tests", *Trans.*, AIME (1953) **198**, 93–102.
4. Arps, J. J. and Roberts, T. G.: "The Effect of the Relative Permeability Ratio, of the Oil Gravity and the Solution Gas-Oil Ratio on the Primary Recovery from a Depletion Type Reservoir", *Trans.*, AIME (1955) **204**, 120–127.

APPENDIX I

Conversion of Units

Theoretical Aspects

Conversion of units of time and length are simple. For example, 1 year (calendar) = 365 days (mean solar) = 525,600 minutes (mean solar) = 3.1536×10^7 seconds (mean solar); and 1 yard = 3 feet = 36 inches = 91.44 centimeters.

The units of force and mass, however, are not as easily converted and understood. The earth exerts a gravitational force on all bodies. The magnitude of this force, called weight, is equal to the mass of the body multiplied by the gravitational acceleration, or

$$F = mg$$

where:

F = weight or force in pounds.
m = mass of the body in slugs.
g = gravitational acceleration, which at sea level is about 32.2 ft/sec^2.

For example, the density (mass per unit volume), ρ, of water having a specific weight, γ, of 62.4 lb/cu ft is

$$\rho = \gamma/g = \frac{62.4 \text{ lb/cu ft}}{32.2 \text{ ft/sec}^2} = 1.94 \text{ slug/cu ft}$$

In the opinion of the writers, it is critical to use different terms for the specific weight, γ, and for the density, ρ.

Example 1. Dynamic Viscosity Conversion Factor

Dynamic or absolute viscosity, μ, may be defined as the (shearing stress)/(rate of shearing strain) ratio assuming a linear distribution of

Conversion of Units

velocity between two plates (one plate moving with respect to the other) with fluid in between:

$$\mu = \frac{F/A}{V/h} \qquad (\text{I-1})$$

$$\mu = \frac{\tau}{V/h} \qquad (\text{I-2})$$

where:

F = the force required to maintain flow (to slide the fluid layers relative to each other, which is accomplished by overcoming the internal fluid friction).
A = area of the moving plate in contact with the fluid.
V = velocity of upper plate if lower plate is stationary.
h = distance between the two plates.
τ = shearing stress or F/A.

The symbols M, L, F, and T represent the fundamental dimensions of mass, length, force, and time, respectively; thus:

$$\mu = \frac{F/L^2}{(L/T)/L} = \frac{FT}{L^2} \qquad (\text{I-3})$$

Inasmuch as force equals mass times acceleration $[F = (ML/T^2)]$, the dimensions of dynamic viscosity are

$$\mu = \frac{(ML/T^2)T}{L^2} = \frac{M}{LT} \qquad (\text{I-4})$$

Thus, dynamic or absolute viscosity can be expressed in lb-sec/ft² or slug/ft-sec, whereas in the metric system:

$$1(\text{dyne-sec/cm}^2) = 1(\text{g/cm-sec}) = 1 \text{ poise}$$

Inasmuch as 1 in. = 2.54 cm and 1 dyne = 2.248×10^{-6} lb,

$$1 \text{ poise} = 1(\text{dyne-sec/cm}^2)$$

$$= \frac{\text{dyne}(2.248 \times 10^{-6} \text{ lb/dyne}) \text{ sec}}{\text{cm}^2[(2.54 \times 12)^2 \text{ ft}^2/\text{cm}^2]}$$

$$= 2.089 \times 10^{-3}(\text{lb-sec/ft}^2)$$
$$= 2.089 \times 10^{-3}(\text{slug/ft-sec})$$

The absolute viscosity of water at 20°C is around 1 centipoise, which is equal to 0.01 of a poise, named in honor of the French scientist Poiseuille.

Example 2. Determination of Multicomponent Conversion Constants

When handling equations using field or laboratory data, it is often easier to develop one constant rather than convert each constituent individually. The following example is the determination of the constant for Darcy's law equation when information is given as follows:

$$q = (\text{constant}) \frac{(k)(\Delta p)(A)}{(\mu)(L)} \tag{I-5}$$

where permeability, k, is in darcys; pressure drop, Δp, is in inches of mercury instead of atm; cross-sectional area, A, is in square inches instead of cm^2; viscosity, μ, is in poises instead of centipoises; length, L, is in yards, instead of cm; and volumetric rate of fluid flow, q, is in quarts per minute instead of cm^3/sec.

From the conversion tables:

1 qt = 946.3529 cu cm	1 sq in. = 6.4516 sq cm
1 min = 60 sec	1 yd = 91.44 cm
1 poise = 100 cp	1 in. Hg = 0.0334211 atm

Thus:

$$q = \frac{(\text{qt})(946.3529 \text{ cm}^3/\text{qt})}{(\text{min})(60 \text{ sec/min})}$$

$$= (\text{constant}) \frac{(\text{in. Hg})(0.0334211 \text{ atm/in. Hg})(\text{in.}^2)(6.4516 \text{ cm}^2/\text{in.}^2)}{(\text{poises})(100 \text{ cp}/1 \text{ poise})(\text{yd})(91.44 \text{ cm/yd})}$$

Solving for the constant gives

$$\text{constant} = \frac{(0.0334211)(6.4516)(60)}{(100)(91.44)(946.3529)}$$

$$= 1.49 \times 10^{-6}$$

If conversion from in. Hg to atmospheres was not given, then the following information should have been given: (a) specific gravity of Hg and (b) 1 atm = 1033 cm of liquid having density of 1.00 g/cm³.

Conversion of Units

CONVERSION FACTORS

Multiply →	by →	to obtain
To Obtain ←	by ←	divide

Acre (ac)	4.3560×10^4	sq ft
	6.272640×10^6	sq in.
	1×10^5	sq links (Gunter's)
	4.046856×10^3	sq m
	1.5625×10^{-3}	sq mi (statute)
	4.840×10^3	sq yd
Acre-foot (ac-ft)	4.3560×10^4	cu ft
	1.233482×10^3	cu m
	1.613333×10^3	cu yd
	3.259×10^5	gal (U.S., liq)
Ares	2.471054×10^{-2}	ac
	1×10^2	sq m
	3.861022×10^{-5}	sq mi
Atmosphere (atm)	1.01325	bars
	7.6×10^1	cm of Hg at 0°C
	1.03326×10^3	cm of H_2O at 4°C
	1.01325×10^6	dynes/sq cm
	3.38995×10^1	ft of H_2O at 39.2°F
	1.46960×10^1	psi
Bar	9.86923×10^{-1}	atm
	7.50062×10^1	cm of Hg (0°C)
	1×10^6	dynes/sq cm
	1.45038×10^1	psi
Barrel (petroleum) (bbl)	5.614583	cu ft
	4.2×10^1	gal (U.S., liq)
	1.589828×10^2	liter (l)
Barrel (U.S., dry) (bbl)	9.69696×10^{-1}	bbl (U.S., liq)
	4.083333	cu ft
	7.056×10^3	cu in.
	1.156271×10^{-1}	cu m
Barrel (U.S., liq) (bbl)	1.03125	bbl (U.S., dry)
	4.210938	cu ft
	7.2765×10^3	cu in.
	1.192405×10^{-1}	cu m
Barrel per hour (oil) (bbl/hr)	9.36×10^{-2}	cu ft/min
	7×10^{-1}	gal/min
British thermal unit (Btu)	2.519958×10^2	cal, g
	1.05435×10^{10}	ergs
	2.50201×10^4	ft-poundals
	7.77649×10^2	ft-lb
	3.92752×10^{-4}	hp-hr
	1.05435×10^3	j
	2.92875×10^{-4}	kw-hr
	1.05435×10^3	w-sec

Multiply →	By →	To obtain
To Obtain ←	By ←	Divide

British thermal unit per minute (Btu/min)	2.51996×10^{-1}	cal, kg/min
	2.35651×10^{-2}	hp
	1.75725×10^{1}	j/sec
	1.75725×10^{1}	w
Buckets (British)	1.818435×10^{4}	cu cm
Bushels (British)	1.032056	bu (U.S.)
	3.63687×10^{4}	cu cm
Bushel (U.S.) (bu)	3.523907×10^{4}	cu cm
	1.244456	cu ft
	3.523808×10^{1}	l
Butts (British)	4.769619×10^{-1}	cu m
Calorie, gram (cal, g)	3.968321×10^{-3}	Btu
	4.184	j
Cental	1×10^{2}	lb
Centare	1	sq m
Centimeter (cm)	3.280840×10^{-2}	ft
	3.937008×10^{-1}	in.
	1×10^{-2}	m
	1×10^{7}	mμ
	6.213712×10^{-6}	mi (statute)
	1.093613×10^{-2}	yd
Centimeter-dyne (cm-dyne)	1.019716×10^{-3}	cm-g
	1.019716×10^{-8}	m-kg
	7.375562×10^{-8}	lb-ft
Centimeter-gram (cm-g)	1×10^{-5}	m-kg
	7.233014×10^{-5}	lb-ft
Centimeter of mercury at 0°C (cm of Hg, 0°C)	1.315789×10^{-2}	atm
	1.33322×10^{4}	dynes/sq cm
	3.937008×10^{-1}	in. of Hg, 0°C
	1.93368×10^{-1}	psi
Centimeter of water (4°C) (cm of H$_2$O)	9.67814×10^{-4}	atm
	9.80638×10^{2}	dynes/sq cm
	1.42229×10^{-2}	psi
Centimeter per second (cm/sec)	1.968504	ft/min
	3.280840×10^{-2}	ft/sec
	3.6×10^{-2}	km/hr
	6×10^{-4}	km/min
	2.236936×10^{-2}	mi/hr
	3.728227×10^{-4}	mi/min

Conversion of Units

Multiply →	by →	to obtain
To Obtain ←	by ←	divide

Centimeter per second per second (cm/sec²)	3.6×10^{-2}	km/(hr-sec)
	2.236936×10^{-2}	mi/(hr-sec)
Centipoise (cp)	1×10^{-2}	g/(cm-sec)
	2.419088	lb/(ft-hr)
	6.71969×10^{-4}	lb/(ft-sec)
	2.089×10^{-5}	$(lb_f\text{-sec})/sq\ ft$
Chain (Gunter's)	6.6×10^{1}	ft
	1.25×10^{-2}	mi (statute)
Circle	3.6×10^{2}	deg
	4×10^{2}	grades
	2.16×10^{4}	min (angular)
	6.283185	radians
Cord (cd)	1.28×10^{2}	cu ft
	3.624573	cu m
Cubic centimeter (cu cm)	3.531467×10^{-5}	cu ft
	6.102374×10^{-2}	cu in.
	1×10^{-6}	cu m
	1.307951×10^{-6}	cu yd
	2.641720×10^{-4}	gal (U.S., liq)
	9.99972×10^{-4}	l
	3.381402×10^{-2}	oz (U.S., liq)
	1.056688×10^{-3}	qt (U.S., liq)
Cubic centimeter per second (cu cm/sec)	2.118880×10^{-3}	cu ft/min
	1.585032×10^{-2}	gal (U.S.)/min
	2.641720×10^{-4}	gal (U.S.)/sec
Cubic foot (cu ft)	2.295684×10^{-5}	ac-ft
	2.831685×10^{4}	cu cm
	7.480520	gal (U.S., liq)
	2.831605×10^{1}	l
	9.575065×10^{2}	oz (U.S., liq)
	2.992208×10^{1}	qt (U.S., liq)
Cubic feet of water (cu ft of H_2O, 39.2°F)	6.24262×10^{1}	lb of H_2O
(cu ft of H_2O, 60°F)	6.23663×10^{1}	lb of H_2O
Cubic feet per hour (cu ft/hr)	2.295684×10^{-5}	ac-ft/hr
	7.865791	cu cm/sec
	7.480520	gal (U.S.)/hr
	2.831605×10^{1}	l/hr
Cubic feet per minute (cu ft/min)	1.377410×10^{-3}	ac-ft/hr
	2.295684×10^{-5}	ac-ft/min
	4.719474×10^{2}	cu cm/sec
	7.480520	gal (U.S.)/min
	4.719342×10^{-1}	l/sec

Multiply →	By →	To Obtain
To Obtain ←	By ←	Divide

Cubic inch (cu in.)	1.638706×10^1	cu cm
	5.787037×10^{-4}	cu ft
	2.143347×10^{-5}	cu yd
	4.329004×10^{-3}	gal (U.S., liq)
	1.638661×10^{-2}	l
	5.541125×10^{-1}	oz (U.S., liq)
	1.731602×10^{-2}	qt (U.S., liq)
Cubic inch of H_2O (cu in. of H_2O, 4°C)	3.61263×10^{-2}	lb H_2O
(cu in. of H_2O, 60°F)	3.60916×10^{-2}	lb H_2O
Cubic meter (cu m)	8.107132×10^{-4}	ac-ft
	1×10^6	cu cm
	3.531467×10^1	cu ft
	6.102374×10^4	cu in.
	1.30795	cu yd
	2.64172×10^2	gal (U.S., liq)
	9.99972×10^2	l
Cubic meter per minute (cu m/min)	2.641721×10^2	gal (U.S., liq)/min
	9.99972×10^2	l/min
Cubic yard (cu yd)	7.645549×10^5	cu cm
	2.7×10^1	cu ft
	4.6656×10^1	cu in.
	7.645549×10^{-1}	cu m
	2.019740×10^2	gal (U.S., liq)
	7.64534×10^2	l
	8.07896×10^2	qt (U.S., liq)
Cubic yard per minute (cu yd/min)	4.5×10^{-1}	cu ft/sec
	3.366234	gal (U.S., liq)/sec
	1.274222×10^1	l/sec
Cubit	4.572×10^1	cm
	1.5	ft
	1.8×10^1	in.
Day (mean solar)	2.4×10^1	hr (mean solar)
	1.44×10^3	min (mean solar)
	8.64×10^4	sec (mean solar)
	2.739726×10^{-3}	yr (calendar)
Decimeter (dm)	1×10^1	cm
Decisteres	1×10^{-1}	cu m
Degree (angular) (deg)	2.7777×10^{-3}	circle
	6.0×10^1	min (angular)
	1.11111×10^{-2}	quadrants
	1.745329×10^{-2}	radian
	3.6×10^3	sec (angular)

Conversion of Units

Multiply → To Obtain ←	by → by ←	to obtain divide
Degree per second (angular) (deg/sec)	1.745329×10^{-2} 1.66666×10^{-1} 2.7777×10^{-3}	radians/sec rpm rps
Dekagram	1×10^{3}	g
Dekameter	3.28084×10^{1} 1×10^{1} 1.093613×10^{1}	ft m yd
Demal	1	g-equiv/cu dm
Drachm (British, liq)	3.551631	cu cm
Dram (avdp) (dr)	4.557292×10^{-1} 2.734375×10^{-1} 1.771845 5.69661×10^{-2} 6.25×10^{-2}	dr (troy or apoth) grains g oz (troy or apoth) oz (avdp)
Dyne	1.573663×10^{-2} 1.019716×10^{-3} 1×10^{-5} 7.233014×10^{-5} 2.248089×10^{-6}	grains g newtons poundals lb
Dyne per centimeter (dyne/cm)	1 1.019716×10^{-3} 1.837185×10^{-4}	erg/sq cm g/cm poundals/in.
Dyne per square centimeter (dyne/sq cm)	9.86923×10^{-7} 1×10^{-6} 7.50062×10^{-5} 1.019745×10^{-3} 4.666451×10^{-4}	atm bars cm of Hg, 0°C cm of H_2O, 4°C poundals/sq in.
Ell	1.143×10^{2} 4.5×10^{1}	cm in.
Erg	9.48451×10^{-11} 2.39006×10^{-8} 9.86923×10^{-7} 3.48529×10^{-11} 1 6.24196×10^{11} 2.373036×10^{-6} 7.37562×10^{-8} 1.019716×10^{-3} 1×10^{-7} 2.777777×10^{-14} 9.86895×10^{-10} 1×10^{-7}	Btu cal, g cu cm-atm cu ft-atm dyne-cm electron v ft-poundals ft-lb g-cm j kw-hr l-atm w-sec

272　　　　　　　　　　*Secondary Recovery and Carbonate Reservoirs*

MULTIPLY →	BY →	TO OBTAIN
To OBTAIN ←	BY ←	DIVIDE

Erg per second (erg/sec)	5.69071×10^{-9}	Btu/min
	1.43403×10^{-6}	cal, g/min
	1	dyne-cm/sec
	4.42537×10^{-6}	ft-lb/min
	1.34102×10^{-10}	hp
	1×10^{-7}	j/sec
	1×10^{-7}	w
Erg-second (erg-sec)	1.50932×10^{26}	Planck's constant
Fathom (fath)	1.8288×10^{2}	cm
	6	ft
	7.2×10^{1}	in.
	1.8288	m
	9.87473×10^{-4}	mi (naut., Int.)
	1.136363×10^{-3}	mi (statute)
	2	yd
Foot (ft)	3.048×10^{1}	cm
	9.99998×10^{-1}	ft (U.S. Survey)
	1.2×10^{1}	in.
	3.048×10^{-1}	m
	1.89393×10^{-4}	mi (statute)
	3.333333×10^{-1}	yd
Foot (U.S. Survey)	1.000002	ft
Foot of air (1 atm, 60°F)	3.6083×10^{-5}	atm
	8.9970×10^{-4}	ft of Hg, 32°F
	1.2244×10^{-3}	ft of H_2O, 60°F
	5.3027×10^{-4}	psi
Foot of mercury (ft of Hg, 32°F)	3.048×10^{1}	cm of Hg, 0°C
	1.36085×10^{1}	ft of H_2O, 60°F
	9.43016×10^{1}	oz/sq in.
	5.89385	psi
Foot per hour (ft/hr)	3.048×10^{1}	cm/hr
	5.08×10^{-1}	cm/min
	8.4666×10^{-3}	cm/sec
	1.66666×10^{-2}	ft/min
	1.645788×10^{-4}	knots (Int.)
	1.89393×10^{-4}	mi/hr
	5.260943×10^{-8}	mi/sec
Foot per minute (ft/min)	3.048×10^{1}	cm/min
	5.08×10^{-1}	cm/sec
	1.6666×10^{-2}	ft/sec
	1.136363×10^{-2}	mi/hr

Conversion of Units

MULTIPLY → To Obtain ←	BY → BY ←	TO OBTAIN DIVIDE
Foot per second (ft/sec)	1.09728×10^5	cm/hr
	1.8288×10^3	cm/min
	3.048×10^1	cm/sec
	6.818182×10^{-1}	mi/hr
	1.136363×10^{-2}	mi/min
Foot per second per second (ft/sec^2)	3.048×10^{-1}	m/sec^2
	6.818182×10^{-1}	mi/(hr-sec)
Firkin (U.S.)	2.857143×10^{-1}	bbl (U.S., liq)
	1.203125	cu ft
	8.326747×10^{-1}	Firkin (British)
	3.406775×10^1	l
Foot-poundal (ft-poundal)	3.99678×10^{-5}	Btu
	1.00717×10^{-2}	cal, g
	4.214011×10^5	erg
	3.1081×10^{-2}	ft-lb
Foot-pound (ft-lb)	1.28593×10^{-3}	Btu
	3.24048×10^{-1}	cal, g
	1.35582×10^7	erg
	3.2174×10^1	ft-poundal
	5.0505×10^{-7}	hp-hr
Foot-pound per hour (ft-lb/hr)	2.14321×10^{-5}	Btu/min
	5.050505×10^{-7}	hp
	3.76616×10^{-4}	w
Foot-pound per minute (ft-lb/min)	2.14321×10^{-5}	Btu/sec
	1.6666×10^{-2}	ft-lb/sec
	3.030303×10^{-5}	hp
	2.2597×10^{-2}	w
Furlong	6.6×10^2	ft
	2.01168×10^2	m
	2.2×10^2	yd
Gallon (U.S., liq) (gal)	3.068883×10^{-6}	ac-ft
	3.174603×10^{-2}	bbl (U.S., liq)
	2.380952×10^{-2}	bbl (petrol, U.S.)
	3.785412×10^3	cu cm
	1.336805×10^{-1}	cu ft
	2.31×10^2	cu in.
	4.951132×10^{-3}	cu yd
	8.326747×10^{-1}	gal (British)
	8.593670×10^{-1}	gal (U.S., dry)
	3.785306	l
	1.28×10^2	oz (U.S., liq)
	8	pt (U.S., liq)
	4	qt (U.S., liq)

Multiply → To Obtain ←	By → By ←	To Obtain Divide
Gallon (U.S., liq) per day (gal/day)	5.570023×10^{-3}	cu ft/hr
Gallon (U.S., liq) per hour (gal/hr)	3.068883×10^{-6} 1.336805×10^{-1} 6.309020×10^{-5} 3.785306	ac-ft/hr cu ft/hr cu m/min l/hr
Gallon (U.S., liq) per minute (gal/min)	2.228×10^{-3} 6.308×10^{-2}	cu ft/sec l/sec
Gamma	1×10^{-6}	g
Geepound	1	slug
Gill (U.S.)	1.182941×10^{2} 7.21875 3.125×10^{-2} 8.326747×10^{-1} 1.182908×10^{-1}	cu cm cu in. gal (U.S., liq) gills (British) l
Grade	2.5×10^{-3} 2.5×10^{-3} 9×10^{-1} 5.4×10^{1} 1.570796×10^{-2}	circles circumference deg (angular) min (angular) radian
Grain	3.239945×10^{-1} 6.3546×10^{1} 6.479891×10^{-2} 2.0833×10^{-3} 6.479891×10^{-8}	carats (metric) dynes g oz (apoth or troy) tons (metric)
Gram per centimeter (g/cm)	9.80665×10^{2} 2.54 6.719690×10^{-2} 5.599742×10^{-3} 1.80166×10^{-1}	dyne/cm g/in. lb/ft lb/in. poundals/in.
Gram per cubic centimeter (g/cu cm)	9.80665×10^{2} 1.16236 3.612729×10^{-2} 8.345404	dyne/cu cm poundals/cu in. lb/cu in. lb/gal (U.S., liq)
Gram per square centimeter (g/sq cm)	9.67841×10^{-4} 9.80665×10^{-4} 7.35559×10^{-2} 9.80665×10^{2} 1.422334×10^{-2}	atm bars cm of Hg, 0°C dyne/sq cm psi
Gram-centimeter per second (g-cm/sec)	9.80665×10^{2} 1.31509×10^{-7} 9.80665×10^{-5}	erg-sec hp w

Conversion of Units

Multiply → To Obtain ←	By → By ←	To Obtain Divide
Gravitational constant	9.80621×10^2	cm/sec^2
(at sea level)	3.21725×10^1	ft/sec^2
Hand	1.016×10^1	cm
	4	in.
Hectare (ha)	2.471054	ac
	1×10^2	ares
	1×10^8	sq cm
	1.076391×10^5	sq ft
	1×10^4	sq m
	3.861022×10^{-3}	sq mi
Hectogram (hg)	1×10^2	g
	2.679229×10^{-1}	lb (apoth or troy)
Hectoliter (hl)	1.00028×10^5	cu cm
	3.531566	cu ft
	2.641794×10^1	gal (U.S., liq)
	1×10^2	l
	3.381497×10^3	oz (U.S., liq)
Hectometer (hm)	1×10^4	cm
	1×10^3	dm
	3.280840×10^2	ft
	1×10^2	m
	1.093613×10^2	yd
Hogshead (hhd)	8.421875	cu ft
	1.4553×10^4	cu in.
	2.384809×10^{-1}	cu m
	6.3×10^1	gal (U.S., liq)
	2.384743×10^2	l
Horsepower boiler (hp boiler)	1.31548×10^1	hp (mechanical)
	1.31495×10^1	hp (electrical)
Horsepower (electric) (hp elec)	7.46×10^9	erg/sec
	1.0004	hp (mechanical)
	7.46×10^{-1}	kw
Horsepower-hours (hp-hr)	2.54614×10^3	Btu
	1.98×10^6	ft-lb
	2.68452×10^6	j
Horsepower (mechanical =	2.54248×10^3	Btu (mean)/hr
550 ft-lb/sec) (hp)	4.24356×10^1	Btu /min
	7.06243×10^{-1}	Btu (mean)/sec
	6.41616×10^5	cal, g/hr
	7.457×10^9	erg/sec
	1.98×10^6	ft-lb/hr

Multiply → To Obtain ←	by → by ←	To Obtain Divide
	3.3×10^4	ft-lb/min
	9.99598×10^{-1}	hp (electrical)
	7.457×10^2	j/sec
	7.457×10^{-1}	kw
Horsepower (metric) (hp metric)	9.85923×10^{-1}	hp (electrical)
	9.8632×10^{-1}	hp (mechanical)
Horsepower (water) (hp water)	1.00006	hp (electrical)
	1.00046	hp (mechanical)
Hour (mean solar) (hr)	4.16666×10^{-2}	days (mean solar)
	6×10^1	min (mean solar)
	3.6×10^3	sec (mean solar)
	3.609856×10^3	sec (sidereal)
	5.952381×10^{-3}	week (mean calendar)
Hour (sidereal)	4.16666×10^{-2}	days (sidereal)
	9.972696×10^{-1}	hr (mean solar)
	6×10^1	min (sidereal)
	3.6×10^3	sec (sidereal)
Hundredweight (short)	4.535924×10^1	kg
	1×10^2	lb (avdp)
Inch (in.)	2.54×10^8	Å
	2.54	cm
	8.33333×10^{-2}	ft
	2.54×10^{-2}	m
	2.77777×10^{-2}	yd
Inch of mercury (in. of Hg, 32°F)	3.34211×10^{-2}	atm
	3.38639×10^{-2}	bars
	3.38639×10^4	dynes/sq cm
	9.2624×10^2	ft of air, 60°F, 1 atm
	1.132957	ft of H_2O, 39.2°F
	7.07262×10^1	psf
Inch of mercury (60°F) (in. of Hg, 60°F)	3.33269×10^{-2}	atm
	3.97685×10^4	dyne/sq cm
	7.05269×10^1	psf
Inch of water (4°C) (in. of H_2O, 4°C)	2.4582×10^{-3}	atm
	2.49082×10^3	dyne/sq cm
	7.35539×10^{-2}	in. of Hg, 32°F
	3.612628×10^{-2}	psi
Inch per hour (in./hr)	1.578282×10^{-5}	mi/hr
Inch per minute (in./min)	1.524×10^2	cm/hr
	5	ft/hr
	9.46969×10^{-4}	mi/hr

Conversion of Units

Multiply → To Obtain ←	By → By ←	To Obtain Divide
Joule (absolute) (j)	9.48451×10^{-4}	Btu
	2.39006×10^{-1}	cal, g
	1×10^{7}	ergs
	2.37304×10^{1}	ft-poundals
	7.37562×10^{-1}	ft-lb
	3.72506×10^{-7}	hp-hr
	9.99835×10^{-1}	j (Int.)
	1	w-sec
Joule (Int.)	1.000165	j (abs.)
Joule per second (absolute) (j/sec)	5.69071×10^{-2}	Btu/min
	1.43403×10^{1}	cal, g/min
	1.34102×10^{-3}	hp (mechanical)
	1	w
Kilderkin (British)	8.182957×10^{4}	cu cm
	4.99355×10^{3}	cu in.
	1.8×10^{1}	gal (British)
Kilogram (kg)	9.80665×10^{5}	dynes
	3.215074×10^{1}	oz (apoth or troy)
	7.093163×10^{1}	poundals
	2.679229	lb (apoth or troy)
	6.852177×10^{-2}	slugs
Kilogram per cubic meter (kg/cu m)	6.242796×10^{-2}	lb/cu ft
	3.612729×10^{-5}	lb/cu in.
Kilogram per square centimeter (kg/sq cm)	9.67841×10^{-1}	atm
	9.80665×10^{-1}	bars
	9.80665×10^{5}	dyne/sq cm
	1.422334×10^{1}	psi
Kilogram-meter (kg-m)	2.34048	cal, g (mean)
	9.80665×10^{7}	ergs
	2.32715×10^{2}	ft-poundals
	7.23301	ft-lb
	3.65304×10^{-6}	hp-hr
Knot (Int.)	5.14444×10^{1}	cm/sec
	6.076115×10^{3}	ft/hr
	1	mi (nautical Int.)/hr
	1.150779	mi (statute)/hr
Last (British)	2.909414×10^{3}	l
League (nautical, Int.)	1.150779	leagues (statute)
League (statute)	2.640×10^{3}	fathoms
	1.584×10^{4}	ft
	4.828032	km
	8.689762×10^{-1}	leagues (nautical, Int.)
	3	mi (statute)

278 Secondary Recovery and Carbonate Reservoirs

MULTIPLY →	BY →	TO OBTAIN
TO OBTAIN ←	BY ←	DIVIDE

Light year (light yr)	6.32795×10^4	astronomical units
	9.46055×10^{12}	km
	5.87851×10^{12}	mi (statute)
Link (Gunter's)	6.6×10^{-1}	ft
	2.01168×10^{-1}	m
	1.25×10^{-4}	mi (statute)
Liter (l)	1.000028×10^3	cu cm
	3.531566×10^{-2}	cu ft
	6.102545×10^1	cu in.
	2.641794×10^{-1}	gal (U.S., liq)
	3.381497×10^1	oz (U.S., liq)
	2.113436	pt (U.S., liq)
	1.056718	qt (U.S., liq)
Liter per minute (l/min)	3.531566×10^{-2}	cu ft/min
	2.641794×10^{-1}	gal (U.S., liq)/min
Liter-atmosphere (l-atm)	9.61045×10^{-2}	Btu
	2.42179×10^1	cal, g
	3.53157×10^{-2}	cu ft-atm
	2.40455×10^3	ft-poundals
	7.47356×10^1	ft-lb
	1.01328×10^2	j
Maxwell	3.335635×10^{-11}	E. S. cgs units
	1	gauss-sq cm
	1	line
	9.9967×10^{-1}	maxwells (Int.)
	1×10^{-8}	v-sec
Meter (m)	1×10^2	cm
	3.280839	ft
	3.937008×10^1	in.
	1×10^{-3}	km
	6.213712×10^{-4}	mi (statute)
	1.093613	yd
Meter per hour (m/hr)	3.280840	ft/hr
	6.213712×10^{-4}	mi (statute)/hr
Meter per minute (m/min)	5.468066×10^{-2}	ft/sec
	3.728227×10^{-2}	mi (statute)/hr
Micron (mu or μ)	1×10^{-4}	cm
	3.28084×10^{-6}	ft
	3.93701×10^{-5}	in.
	1×10^{-6}	m
Mile (nautical, British)	1.151515	mi (statute)
Mile (nautical, Int.)	1.150779	mi (statute)

Conversion of Units

Multiply →	By →	To Obtain
To Obtain ←	By ←	Divide

Mile (statute) (mi)	1.609344×10^5	cm
	8×10^1	chains (Gunter's)
	5.28×10^3	ft
	6.336×10^4	in.
	1.609344	km
	1.70111×10^{-13}	light yr
	1.609344×10^3	m
	8.68421×10^{-1}	mi (nautical, British)
	8.689762×10^{-1}	mi (nautical, Int.)
	1.76×10^3	yd
Mile per hour (mi/hr)	4.4704×10^1	cm/sec
	8.8×10^1	ft/min
	1.46666	ft/sec
	2.68224×10^1	m/min
Millibar	1×10^{-3}	bars
Milligram	1×10^{-3}	g
Milliliter (ml)	1.000028	cu cm
	6.102545×10^{-2}	cu in.
	1×10^{-3}	l
	3.381497×10^{-2}	oz (U.S., liq)
Millimeter (mm)	1×10^{-1}	cm
	3.28084×10^{-3}	ft
	3.937008×10^{-2}	in.
	1×10^{-3}	m
Millimeter of mercury (0°C)	1.315789×10^{-3}	atm
(mm of Hg, 0°C)	1.33322×10^{-3}	bars
	1.93368×10^{-2}	psi
Minim (U.S.)	6.161152×10^{-2}	cu cm
Minute (angular)	1.666666×10^{-2}	deg (angular)
	1.85185×10^{-4}	quadrants
	2.908882×10^{-4}	radians
	6×10^1	sec (angular)
Minute (mean solar) (min)	6.944444×10^{-4}	days (mean solar)
	1.666666×10^{-2}	hr (mean solar)
	1.002738	min (sidereal)
Minute (angular) per centimeter (min, angular/cm)	2.908882×10^{-4}	radians/cm
Month (mean calendar) (mo)	3.041667×10^1	days (mean solar)
	7.3×10^2	hr (mean solar)
	1.030005	mo (lunar)
	4.345238	week (mean calendar)
	8.333333×10^{-2}	yr (calendar)

| MULTIPLY → | BY → | TO OBTAIN |
| To OBTAIN ← | BY ← | DIVIDE |

Myriagram	1×10^4	g
	2.204623×10^1	lb (avdp)
Newton	1×10^5	dynes
	2.248089×10^{-1}	lb
Newton-meter	1×10^7	dyne-cm
	7.375621×10^{-1}	lb-ft
Noggin (British)	1.420652×10^2	cu cm
	3.125×10^{-2}	gal (British)
Ounce (apoth or troy) (oz)	4.80×10^2	grains
	3.110349×10^1	g
	1.097143	oz (avdp)
	8.333333×10^{-2}	lb (apoth or troy)
	3.428571×10^{-5}	tons (short)
Ounce (avdp) (oz)	9.114583×10^{-1}	oz (apoth or troy)
Pace	7.62×10^1	cm
	2.5	ft
	3×10^1	in.
Palm	7.62	cm
	2.5×10^{-1}	ft
	3	in.
Parsec	3.08374×10^{13}	km
	1.91615×10^{13}	mi (statute)
Part per million (ppm)	1×10^{-3}	g/l
Peck (U.S.) (pk)	8.809767×10^3	cu cm
	3.111140×10^{-1}	cu ft
	5.37605×10^2	cu in.
	2.327294	gal (U.S., liq)
	8.809521	l
Pennyweight (dwt)	1.555174	g
	4.1666×10^{-3}	lb (apoth or troy)
Perch (masonry)	2.475×10^1	cu ft
Pica (printer's)	4.217518×10^{-1}	cm
	1.66044×10^{-1}	in.
Pint (U.S., liq) (pt)	4.731765×10^2	cu cm
	1.671007×10^{-2}	cu ft
	2.8875×10^1	cu in.
	6.188915×10^{-4}	cu yd
	1.25×10^{-1}	gal (U.S., liq)
	4.731632×10^{-1}	l
	1.6×10^1	oz (U.S., liq)
	5×10^{-1}	qt (U.S., liq)

Conversion of Units

Multiply → To Obtain ←	By → By ←	To Obtain Divide
Planck's constant	6.6255×10^{-27}	erg-sec
Point (printer's)	3.514598×10^{-2}	cm
Poise	1	g/cm-sec
	3.6×10^{2}	kg/m-hr
	6.72×10^{-2}	lb/ft-sec
	1.45×10^{-5}	Reyn (lb_f-sec/in.^{2})
Pottle (British)	5×10^{-1}	gal (British)
	2.27298	l
Poundal	1.38255×10^{4}	dynes
	1.409808×10^{1}	g
	3.1081×10^{-2}	lb (avdp)
Pound (apoth or troy) (lb)	3.732417×10^{2}	g
	1.2×10^{1}	oz (apoth or troy)
	8.228571×10^{-1}	lb (avdp)
	3.673469×10^{-4}	tons (long)
	3.732417×10^{-4}	tons (metric)
	4.114286×10^{-4}	tons (short)
Pound (avdp) (lb)	1.215277	lb (apoth or troy)
Puncheon (British)	3.179751×10^{-1}	cu m
	8.4×10^{1}	gal (U.S., liq)
Quadrant	5.4×10^{3}	min (angular)
	1.570796	radians
Quartern (British, dry)	2.273044×10^{3}	cu cm
Quartern (British, liq)	1.420652×10^{2}	cu cm
Quarter (U.S., long)	2.540117×10^{2}	kg
	5.6×10^{2}	lb (avdp)
Quarter (U.S., short)	2.267962×10^{2}	kg
	5×10^{2}	lb
Quart (British) (qt)	1.136522×10^{3}	cu cm
	1.200949	qt (U.S., liq)
Quart (U.S., dry) (qt)	1.101221×10^{3}	cu cm
	6.720062×10^{1}	cu in.
	2.909118×10^{-1}	gal (U.S., liq)

Multiply → To Obtain ←	By → By ←	To Obtain Divide
Quart (U.S., liq) (qt)	9.463530 × 10²	cu cm
	3.342014 × 10⁻²	cu ft
	5.775 × 10¹	cu in.
	2.5 × 10⁻¹	gal (U.S., liq)
	3.2 × 10¹	oz (U.S., liq)
	2	pt (U.S., liq)
	8.326747 × 10⁻¹	qt (British)
	8.59367 × 10⁻¹	qt (U.S., dry)
Quintal (metric)	1 × 10⁵	g
	2.204623 × 10²	lb (avdp)
Radian	1.591549 × 10⁻¹	circumference
	5.729578 × 10¹	deg (angular)
	3.437747 × 10³	min (angular)
	6.366198 × 10⁻¹	quadrants
	1.591549 × 10⁻¹	revolutions
	2.062648 × 10⁵	sec (angular)
Radian per centimeter	1.746375 × 10³	deg (angular)/ft
	1.455313 × 10²	deg (angular)/in.
Radian per second (radian/sec)	5.729578 × 10¹	deg (angular)/sec
	9.549297	rpm
	1.591549 × 10⁻¹	rps
Radian per second per second (radian/sec²)	5.729578 × 10²	revolution/min²
	9.549297	revolution/min-sec
Register ton	1 × 10²	cu ft
	2.831685	cu m
Revolution (angular)	3.6 × 10²	deg (angular)
	4 × 10²	grades
	4	quadrants
	6.283185	radians
Reyn	6.89476 × 10⁶	centipoises
Rhes	1	poise⁻¹
Rod	5.0292 × 10²	cm
	1.65 × 10¹	ft
	1.98 × 10²	in.
	5.0292	m
	3.125 × 10⁻³	mi (statute)
Rod (British, volume)	1 × 10³	cu ft
	2.831685 × 10¹	cu m
Rood (British)	2.5 × 10⁻¹	ac
	1.21 × 10³	sq yd

Conversion of Units

Multiply → To Obtain ←	By → By ←	To Obtain Divide
Rope (British)	2×10^1	ft
	6.096	m
	6.666666	yd
Scruple (apoth)	4.1666×10^{-2}	oz (apoth or troy)
Seam (British)	1.027479×10^1	cu ft
	2.909414×10^2	l
Second (angular) (sec)	2.77777×10^{-4}	deg (angular)
	1.66666×10^{-2}	min (angular)
	4.848137×10^{-6}	radians
Second (mean solar)	1.157407×10^{-5}	days (mean solar)
	2.777777×10^{-4}	hr (mean solar)
	1.666666×10^{-2}	min (mean solar)
	1.002738	sec (sidereal)
Skein	3.6×10^2	ft
	1.09728×10^2	m
Slug	1.45939×10^1	kg
	3.21740×10^1	lb (avdp)
Slug per cubic foot (slug/cu ft)	5.15379×10^{-1}	g/cu cm
Space (entire)	2	hemispheres
	1.256637×10^1	steradians
Span	2.286×10^1	cm
	7.5×10^{-1}	ft
	9	in.
Spherical right angle	2.5×10^{-1}	hemispheres
	1.25×10^{-1}	spheres
	1.570796	steradians
Square centimeter (sq cm)	1.076391×10^{-3}	sq ft
	1.550003×10^{-1}	sq in.
	1×10^{-4}	sq m
	1.19599×10^{-4}	sq yd
Square chain (Gunter's)	1×10^{-1}	ac
	4.356×10^3	sq ft
	6.27264×10^5	sq in.
	4.046856×10^2	sq m
	1.5625×10^{-4}	sq mi
	4.84×10^2	sq yd
Square degree	3.046174×10^{-4}	steradians

Multiply → To Obtain ←	By → By ←	To Obtain Divide
Square foot (sq ft)	2.295684×10^{-5}	ac
	9.290304×10^{2}	sq cm
	1.44×10^{2}	sq in.
	3.587006×10^{-8}	sq mi
	1.111111×10^{-1}	sq yd
Square inch (sq in.)	6.4516	sq cm
	6.9444×10^{-3}	sq ft
	2.490977×10^{-10}	sq mi
Square link (Gunter's) (sq link)	1×10^{-5}	ac
	4.046856×10^{2}	sq cm
	4.356×10^{-1}	sq ft
	6.27264×10^{1}	sq in.
Square meter (sq m)	2.471054×10^{-4}	ac
	1×10^{4}	sq cm
	1.076391×10^{1}	sq ft
	1.550003×10^{3}	sq in.
	3.861022×10^{-7}	sq mi
	1.19599	sq yd
Square mile (sq mi)	6.4×10^{2}	ac
	2.787829×10^{7}	sq ft
	2.589988×10^{6}	sq m
	3.0976×10^{6}	sq yd
Square rod (sq rod)	6.25×10^{-3}	ac
	2.529285×10^{5}	sq cm
	2.7225×10^{2}	sq ft
	3.9204×10^{4}	sq in.
	9.76562×10^{-6}	sq mi
	3.025×10^{1}	sq yd
Square yard (sq yd)	2.066116×10^{-4}	ac
	8.361274×10^{3}	sq cm
	9	sq ft
	1.296×10^{3}	sq in.
	8.361274×10^{-1}	sq m
	3.228306×10^{-7}	sq mi
Steradian	1.591549×10^{-1}	hemispheres
	7.957747×10^{-2}	solid angles
	7.957747×10^{-2}	spheres
	6.366198×10^{-1}	spher. right angle
	3.282806×10^{3}	sq deg
Stere	1	cu m
	9.99972×10^{2}	l

Conversion of Units

Multiply → To Obtain ←	By → By ←	To Obtain Divide
Stoke	1	sq cm/sec
	1.550003×10^{-1}	sq in./sec
	1	poise-cu cm/g
Ton (long)	1.016047×10^{3}	kg
	2.240×10^{3}	lb (avdp)
	1.12	tons (short)
Ton (metric)	1×10^{3}	kg
	2.204623×10^{3}	lb (avdp)
	9.842065×10^{-1}	tons (long)
	1.102311	tons (short)
Ton (short)	8.89644×10^{8}	dynes
	9.071847×10^{2}	kg
	3.2×10^{4}	oz (avdp)
	2.430555×10^{3}	lb (apoth or troy)
	2×10^{3}	lb (avdp)
	8.928571×10^{-1}	tons (long)
	9.071847×10^{-1}	tons (metric)
Ton (long) per square foot (ton, long/sq ft)	1.05849	atm
	1.093664×10^{3}	g/sq cm
	2.24×10^{3}	psf
Ton (short) per square foot (ton, short/sq ft)	9.45082×10^{-1}	atm
	9.57605×10^{2}	dynes/sq cm
	9.76486×10^{2}	g/sq cm
	1.38888×10^{1}	psi
Township (U.S.)	2.304×10^{4}	ac
	3.6×10^{1}	sections
	3.6×10^{1}	sq mi
Tun	2.52×10^{2}	gal (U.S., liq)
Watt (w)	3.41443	Btu/hr
	8.60421×10^{2}	cal, g/hr
	1×10^{7}	erg/sec
	4.42537×10^{1}	ft-lb/min
	1.34102×10^{-3}	hp
	1	j/sec
	1×10^{-3}	kw
Watt (Int.) (w)	1.000165	w
Watt-hour (w-hr)	3.41443	Btu
	3.40952	Btu (mean)
	8.60421×10^{2}	cal, g
	3.6×10^{3}	j

Multiply	→	By	→	To Obtain
To Obtain	←	By	←	Divide

Week (mean calendar)	7	days (mean solar)
	1.68×10^2	hr (mean solar)
	1.008×10^4	min (mean solar)
	2.30137×10^{-1}	months (mean calendar)
	1.917808×10^{-2}	yr (calendar)
Wey (British, mass)	2.52×10^2	lb (avdp)
Yard (yd)	9.144×10^1	cm
	3	ft
	3.6×10^1	in.
	9.144×10^{-1}	m
Year (calendar) (yr_c)	3.65×10^2	days (mean solar)
	8.76×10^3	hr (mean solar)
	5.256×10^5	min (mean solar)
	1.236006×10^1	months (lunar)
	1.2×10^1	months (mean calendar)
	3.1536×10^7	sec (mean solar)
	5.214286×10^1	weeks (mean calendar)
	9.992981×10^{-1}	yr (sidereal)
	9.993369×10^{-1}	yr (tropical)
Year (leap) (yr_l)	3.66×10^2	days (mean solar)
Year (sidereal) (yr_s)	1.000702	yr (calendar)
Year (tropical) (yr_t)	1.000663	yr (calendar)

Temperature Conversion Formulas

To obtain	Formula
°F (Fahrenheit)	$(°C \times 1.8) + 32$
°C (Centigrade)*	$\dfrac{°F + 40}{1.8} - 40$
°C	$(°F - 32) \times 0.5555$
°K (Kelvin)	$°C + 273.16$
°R (Rankine)	$°F + 459.688$

* Or Celcius, which is preferred for international use.

Bibliography

1. A.I.S.C.: *Steel Construction Manual*, American Institute of Steel Construction, New York (1958) 420 pp.
2. Binder, R. C.: *Fluid Mechanics*, 4th ed., Prentice-Hall, Englewood Cliffs, New Jersey (1962) 453 pp.
3. Burington, R. S.: *Handbook of Mathematical Tables and Formulas*, Handbook Publishers, Sandusky, Ohio (1957) 296 pp.
4. Chilingar, G. V. and Beeson, C. M.: *Surface Operations in Petroleum Production*, Elsevier, New York (1969) 397 pp.
5. Frick, T. C. (Editor): *Petroleum Production Handbook*, McGraw-Hill, New York (1962) Vols. 1 and 2.
6. Weast, R. C. (Editor): *Handbook of Chemistry and Physics*, 49th Ed., Chemical Rubber Co., Cleveland, Ohio (1968) 2074 pp.
7. Zaba, J. and Doherty, W. T.: *Practical Petroleum Engineer's Handbook*, 4th Ed., Gulf Publishing Co., Houston, Texas (1956) 818 pp.

Author Index

Abernathy, B. F., 45, 46, 48–50, 52–53, 55, 97, 101, 102, 103, 104, 108, 110, 114, *127*
Ache, P. S., 45, 46, 48–50, 52–53, *96*
Agan, J. B., 159, *165*
Akins, D. W., Jr., 112, 114, *127*
Allen, D. R., 222, *224*
Allen, E. E., 61, 62–65, *98*
Allen, H. H., 126, *129*
Allen, W. W., 114, *128*
Amyx, J. W., 30, *43*, 186, 187, *195*
Anderson, R. C., 110, *127*
Andresen, K. H., 92, *98*
Andriasov, R., 181, *182*
Archer, D. L., 121, 122, *128*
Armstrong, T. A., 91, *98*
Arnold, C. W., 155, *164*
Aronofsky, J. S., 45, 46, 48–50, 52–53, *96*
Arps, J. J., 45, 46, 48–50, 52–53, *96*, 123–125, *129*, 252, *263*

Babalyan, G. A., 94n, 226, 227, 228, 229, 230, *242*
Baker, J. D., 45, 46, 48–50, 52–53, 56, 61, *96*
Barfield, E. C., 120, *128*
Bass, D. M., Jr., 30, *43*, 186, 187, *195*
Bayazeed, A. F., 139, 140, *144*
Beal, C., 197, 198, 203, *206*
Beeson, C. M., *252*, *287*
Beeson, D. M., 17, *19*
Benham, A. L., 146, 147, 151–162, *164*
Bernard, G. C., 230, *242*
Berry, V. J., Jr., 139, 141, *144*
Binder, R. C., 173, 174, 175, *182*, *287*
Black, J. L., 123, *129*
Blair, P. M., 45, 46, 48–50, 52–53, *97*
Blanton, J. R., 159, *165*
Bleakley, W. B., 51, *98*, 110, 112, *127*, 159, *165*
Bohannon, D. L., 139, *144*
Boley, D. W., 45, 46, 48–50, 52–53, 55, *97*
Borgan, R. L., 115, *128*
Bridges, P. M., 45, 46, 55, *96*
Brinkman, F. H., 154, *164*

Brownscombe, E. R., 119, *128*
Bruce, W. A., *252*
Buckley, S. E., 39, *44*, 45, 46, 48–50, 52–53, 55, 72, 74–83, 90, 92, *96*, 99, 100, 104, 107, 110, 131, 132
Buckwalter, J. H., 41, *44*
Burcik, E. J., 91, *98*
Burdine, N. T., 31, 32, *43*
Burington, R. S., *287*
Burrows, D. B., 197, 200, 201, 202, *206*

Calhoun, J. C., 45, 46, 48–50, 52–53, 54, 96, 100, *127*
Callaway, F. H., 51, 53, *98*
Cargile, L. L., 123, *128*
Carman, P. C., 244, *252*
Carr, N. L., 197, 200, 201, 202, *206*
Caudle, B. H., 45, 46, 48–50, 52–53, 55, 66–68, *96*, *98*, 108, 109, 110, *127*
Chalkley, H. W., 243, 249, *252*
Chew, J., 203, *206*
Chierici, G., 185, 186, 187, 188, 189, *195*, 222, *224*
Chilingar, G. V., 171, *172*, 220–221, *224*, 230, 231, 242, 246, *252*, 253, *263*, *287*
Chilingarian, G. V. (see Chilingar, G. V.), 210, *224*
Clark, N. J., 145, 146, 147, 160, *164*, *165*
Clay, T. W., 139, *144*
Coats, K. H., 39, 40, *44*, 159, *165*
Cobb, T. R., 45, 46, 48–50, 52–53, *96*
Cook, A. B., 139, 140, *144*
Cooke, J. T., 130, *143*
Corey, A. T., 32, 34, 35, 36, 37, 38, *44*
Cornell, D., 139, *144*, 245, *252*
Cornfield, J., 242, 249, *252*
Cosgrove, J. J., 45, 46, 48–50, *97*
Craft, B. C., 216, *224*, 247, *252*
Craig, F. F., Jr., 45, 46, 48–50, 52–53, 55, 70, *97*, 101, 102, 103, 104, 105, 106, 108, 110, *127*, 155, *164*, 253, *263*
Crawford, P. B., 69–70, 71, 72, *98*, 150
Craze, R. C., 14, *18*, 29, *43*
Crosby, G. E., 147, *164*

289

Davison, K., 160, *165*
de Nevers, N. H., 17, *19*
Deppe, J. C., 45, 46, 48–50, 56, 57–60, *96*
DeWitt, S. N., 115, *128*
Dickey, P. A., 92, *98*
Dobrynin, V. M., 222, 223, *224*
Dodson, C. R., 212, 218, 219, *224*
Doepel, G. W., 159, *165*
Doherty, W. T., *287*
Donahoe, C. W., 139, *144*
Douglas, J., 45, 46, 48–50, 52–53, *97*
Dowden, W. E., 146, 147, 151–162, *164*
Downie, J., 159, *164*
Dyes, A. B., 45, 46, 48–50, 52–53, 55, 66–68, *98*, 110, 119, *127*, *128*
Dykstra, H., 39, *44*, 45, 46, 48–50, 52–53, 54, 74, 84–87, *95*, 99, 100, 110, 153

Earlougher, R. C., 45, 46, 47, 48–50, 92, *97*
Elenbaas, J. R., 139, *144*
Elkins, L. E., 130, 135, 142, 143, *144*
Elkins, L. F., 116, 117, 121, *128*
Enright, R. J., 145, *164*
Eremenko, N. A., 247, *252*
Erickson, R. A., 45, 46, 48–50, 52–53, 55, *96*, 110, *127*
Essley, P. L., Jr., 3, 13, *18*
Ewing, S. P., 154, *164*

Fahhad, S. A., 220–222, *224*
Faris, S. R., 246, *252*
Fatt, I., 219, *224*
Felsenthal, M., 45, 46, 48–50, 52–53, 55, *96*; 122, *128*
Fernandes, R. J., 159, *165*
Ferrell, H. H., 122, *128*
Fickert, W. E., 115, *128*
Fisher-Rosenbaum, M. J., 40, *44*
Fitch, R. A., 159, 160, *165*
Frank, J. R., 115, *128*
Frick, T. C., *287*
From, D. T., 160, *165*

Garder, A. O., Jr., 159, *164*
Gardner, G. H., 30, 33, 34, *43*, 159, *164*
Garms, K., 145, 146, 147, *164*
Gealy, J. D., Jr., 114, *127*
Geertsma, J., 222, *224*
Geffen, T. M., 45, 46, 48–50, 52–53, 55, 70, *97*, 101, 102, 104, 105, 106, 110, *127*

Ghose, S. K., 220–221, *224*
Gimatudinov, Sh., 181, *182*
Gogarty, W. B., 91, *98*
Goolsby, J. L., 14, *18*, 52, *98*, 100, *127*
Gould, R. C., 121, *128*
Gournay, L. B., 246, *252*
Govorova, G., 181, *182*
Graham, J. W., 117, 118, 119, *128*
Grant, H. K., 15, *18*
Graue, D. J., 204, 205, *206*
Graves, E. D., Jr., 247, *252*
Greenberger, M. H., 45, 46, *97*
Greenkorn, R. A., 159, *164*
Greer, F. C., *252*
Gregory, A. R., 246, *252*
Griffith, J. D., 159, *165*
Guerrero, E. T., 45, 46, 47, 48–50, 92, *97*
Guidroz, G. M., 120–121, *128*
Guthrie, R. K., 45, 46, *97*

Habermann, B., 70, 98, 102, 110, *127*, 148, 149, 150, 157, 158, *164*
Hall, H. N., 45, 46, 55, *96*, 219, *224*
Halloway, H. D., 160, *165*
Haring, R. E., 159, *164*
Harvey, R. D., 17, *19*
Hauber, W. C., 45, 46, 48–50, *96*
Hawkins, M. F., 216, *224*
Hazebroek, P., 61, 62–65, *98*
Henderson, J. H., 45, 46, 48–50, *97*
Hendrickson, G. E., 45, 46, 48–50, *97*, 101, 104, 110, 114, *127*
Henry, J. C., 100–101, *127*
Herbeck, E. F., 159, *165*
Herriot, H. P., 114, *128*
Hester, C. T., 122, *128*
Heuer, G. J., 45, 46, 48–50, 52–53, *96*
Hewitt, C. H., 13, *18*
Hiatt, W. N., 45, 46, 48–50, *97*
Higgins, R. V., 45, 46, 48–50, 52–53, 55, *97*
Hnatiuk, J., 142, *144*
Holden, W. R., 247, *252*
Holm, L. W., 16, *18*
Hossain, A. K. M. S., 231–235, 238, 240, *242*
Hovanessian, S. A., 77, *98*
Hurst, W., 45, 46, 48–50, 52–53, 55, *96*

Inks, C. G., 94, *98*

Author Index

Jackson, R. W., 15, *18*
Jacoby, R. H., 139, 141, *144*
Jewett, R. L., 45, 46, 48–50, 52–53, 56, 61, *96*
Johnson, C. A., 159, *164*
Johnson, C. E., Jr., 34–38, *44*, 45, 46, 48–50, 52–53, 81, 85, 86, 88, 89, 92, 93, *95*, *98*
Johnson, E. F., 154, *164*
Johnson, F. S., 139, 140, *144*
Jordan, J. K., 120, *128*

Katz, D. L., 139, *144*, 183, 184, 187, 194, *195*, 245, *252*
Kaufman, A., 4–5, 6–7, 8–9, 10–12, *18*
Keller, W. O., 14, *18*
Kemp, C. E., 66–68, *98*
Kendall, H. A., 159, *164*
Khanin, A. A., 169, 170, *172*, 250–251, *252*
Knopp, C. R., 187, *195*
Kobayashi, R., 197, 200, 201, 202, *206*
Kock, H. A., 150, 157, *164*
Kotyakhov, F. I., 166, 169, *172*, 246, 247, *252*
Koval, E. J., 159, *164*
Kovalenko, E. K., 94n
Kunzman, W. J., 146, 147, 157–162, *164*

Lacey, J. W., 163, *165*
Lacik, H. A., 123, *129*
Lahring, R. I., 94, *98*
Landrum, B. L., 69–70, 71, 72, *98*
Langton, J. R., 159, *165*
Larson, V. C., 163, *165*
Lasater, J. A., 185, 187, 191, *195*
Leighton, A. J., 45, 46, 48–50, 52–53, 55, *97*
Leverett, M. C., 39, *44*, 45, 46, 48–50, 52–53, 55, 72, 74–83, 90, 92, *96*, 99, 100, 104, 110, 131, 132
Lewis, W. B., 72, *98*
Lindner, J. D., 160, *165*
Lipson, L. B., 246, *252*
Locker, G. R., 142, *144*
Lohec, R. E., 139, *144*
Long, G., 185, 186, 187, 188, 189, *195*, 222, *224*
Longren, H. F., 141, *144*
Luffel, D. L., 155, *164*

Mahaffey, J. L., 148, 158, *164*
Main, R., 246, *252*
Mannon, R. W., 253, *263*
Marchant, L. C., 17, *19*
Marrs, D. G., 161, 162, *165*
Marsden, S. S., 173, 177, *182*, 232, 234, 236, 237, 239, 241, *242*
Martin, J. C., 211, *224*
Martinelli, J. W., 142, *144*
Masson, P. H., 246, *252*
Matthews, C. S., 40, 44, 45, 46, 48–50, 52–53, 56, 61, *96*, 148, 158, *164*
McCaleb, J. A., 115, 117, *128*
McCarthy, J. C., 100, *127*
McCulloch, R. C., 159, *165*
McGraw, J. H., 139, *144*
McShane, J. B., Jr., 115, *128*
Menzie, D. E., 17, *19*
Miller, F. H., 51, *98*, 114, *127*.
Montgomery, E. F., III, 231–235, 238, 240, *242*
Moore, J. L., 145, 146, 147, *164*
Moore, W. D., 120, *128*
Morel-Seytoux, H. J., 45, 46, 48–50, *97*
Moring, J. D., 100–101, *127*
Morse, R. A., 14, *18*, 45, 46, 48–50, 52–53, 55, 70, *97*, 100, 101, 102, 104, 105, 106, 110, *127*, 253, *263*
Mungan, N., 91, *98*
Muravyov, I., 181, *182*
Muskat, M., 45, 46, 48–50, 52–53, 55, 56, *96*, 108, *127*, 130, *143*, 166, 167, *172*

Naar, J., 45, 46, 48–50, *97*
Neslage, F. J., 17, 18, 141, *144*
Nolan, W. E., 142, *144*

O'Briant, J. F., 114, *127*
Odeh, A. S., 39, 40, *44*
Oefelein, F. H., 41–42, *44*
Owens, W. W., 121, 122, *128*, 156, *164*

Park, H., 243, 249, *252*
Park, R. A., 110, 111, *127*
Parsons, R. L., 39, *44*, 45, 46, 48–50, 52–53, 54, 74, 84–87, *95*, 99, 100, 110, 153
Peaceman, D. W., 45, 46, 48–50, *97*, 159, *164*
Perez-Rosales, C., 248, 249, 251, *252*
Perkins, A., 51, *98*, 114, *127*

Peterson, R. B., 163, *165*
Pirson, S. J., 247, *252*
Poettmann, F. H., 139, *144*
Polozkov, V., 181, *182*
Poston, S. W., 231–235, 238, 240, *242*
Pozzi, A. L., Jr., 159, *164*
Prats, M., 45, 46, 48–50, 52–53, 56, 61, 62–65, 96, *98*
Purcell, W. R., 30, 31, *43*, 246, *252*
Pye, D. J., 91, *98*

Rachford, H. H., 45, 46, 48–50, *97*, 155, *164*
Rahme, H. S., 45, 46, 48–50, 52–53, *96*
Ramey, H. J., Jr., 45, 46, 48–50, 52–53, 77, 80, *96*, *98*, 207–216, *223*, 231–235, 237–241, *242*
Ramsey, L. A., 187, *195*
Rathjens, C. H., 34, *44*
Read, D. L., 147, *164*
Richardson, J. G., 117, 118, 119, *128*
Rieke, H. H., III, 220–221, *224*, 230, 231, *242*, 253, *263*
Riki, K., 139, *144*
Roberts, G. R., Jr., 2, 4, 16, *18*
Roberts, T. G., 45, 46, 48–50, 52–53, 55, 77, 78, 80, *97*, 160, *165*, 253, *263*
Robertson, J. O., Jr., 41, 42, *44*, 220, *224*, 230, 231, *242*
Rose, W. D., 246, *252*
Rowan, G., 14, *18*
Rowe, A. M., Jr., 159, *164*
Rutherford, W. M., 148, 158, *164*

Saniford, B. B., 91, *98*
Sawabini, C. T., 219, 220, 221, *224*
Sawyer, G. H., 122, *128*
Schauer, P. E., 45, 46, 48–50, *97*
Schmalz, J. P., 45, 46, 48–50, 52–53, *96*
Schoeppel, R. J., 45, 46, 47, 48–50, 52, 53, 54, 55, *97*
Schultz, W. P., 145, 146, 147, *164*
Schurz, G. F., 92, *98*
Scott, E. Z., 147, *164*
Sessions, R. E., 158, 161, *164*
Shearin, H. M., Jr., 145, 146, 147, *164*, 246, *252*
Shirkovskiy, A. I., 248, *252*
Sibley, W. P., 159, *165*
Silberberg, I. H., 159, *164*
Simon, R., 204, 205, *206*

Simpson, R. E., 253–*263*
Sinnokrot, A., 232, 234, 236, 237, 239, 241, *242*, 246, *252*
Skov, A. M., 121, *128*
Slider, H. C., 45, 46, 48–50, 52–53, *96*
Slobod, R. L., 45, 46, 48–50, 52–53, 55, *96*, 110, *127*, 150, 157, *164*
Smith, F. W., 91, *98*
Smith, J. W., 252–*262*
Smith, R. C., 17, *19*, 40, *44*, 153, *164*
Snell, G. W., 92, *98*
Snyder, R. W., 77, 80, *98*
Spangler, M. R., 32, *43*, 246, *252*
Spencer, G. B., 139, 140, *144*
Spivak, A., 159, *165*
Standing, M. B., 183, 184, 185, 186, 190, 192, 193, *195*, 210, 212, 215, 217, 218, 219, *223*, *224*
Steward, C. R., 253–*263*
Stiehler, R. D., 114, *128*
Stiles, W. E., 45, 46, 48–50, 52–53, 74, 87–90, *96*, 99, 100, 101, 102, 103, 104, 108, 110, 123, *127*, 153
Stone, H. L., 155, *164*
Sturdivant, W. C., 162, 163, *165*
Suder, F. E., 45, 46, 48–50, 52–53, 54, *96*

Talkington, G. E., 115, *128*
Tarner, J., 39, *44*, 131, 135–138, *143*
Terwilliger, P. L., 45, 46, 55, *96*
Thomas, J. B., 126, *129*
Thompson, J. L., 91, *98*
Tittle, R. M., 160, *165*
Tkhostov, B. A., 222–223, *224*
Torrey, P. D., 13, 14, *18*, 20, 28, *43*, 92, *98*
Trube, A. S., Jr., 115, *128*, 222, *224*

van der Knaap, W., 219, 222, *224*
van der Vlis, A. C., 219, *224*
van Wingen, N., 45, 197, 199, *206*
Vary, J. A., 139, *144*
Vendelshteyn, B. Yu., 222, 223, *224*
Vennard, J. K., 173, 174, 175, *182*
Vezirova, A. D., 222, 223, *224*

Wagner, R. J., 45, 46, 48–50, 52–53, *97*
Walker, J. W., 123, *128*
Walker, S. W., 2, 4, 16, *18*
Warren, J. E., 14, *18*, 45, 46, 48–50, *97*

Author Index

Wayhan, D. A., 115, *128*
Weast, R. C., *287*
Webb, T. S., 246, *252*
Weinany, C. F., 139, *144*
Welge, H. J., 45, 46, 48–50, 52–53, 55, *97*, 110, 131–135, *143*, 154, *164*
Welsh, J. R., 253–*263*
Whitting, R. L., 30, *43*, 186, 187, *195*
Williams, M., 246, *252*
Willingham, R. W., 117, *128*
Willman, G. J., 162, *165*
Wilsey, L. E., 45, 46, 55, *96*
Wilson, J. F., 112, *127*, 159, *164*
Wilson, W. W., 13, *18*

Witte, M. D., 45, 46, 48–50, 52–53, 55, *96*, 108, 109, *127*
Winsauer, W. O., 246, *252*
Wright, F. F., 15, *18*
Wood, B. O., 115, *128*
Wyllie, M. R. J., 30, 32, 33, 34, *43*, 245, 246, *252*

Ysrael, S., 231–235, 238, 240, *242*
Yust, C. S., 253–*263*
Yuster, S. T., 45, 46, 48–50, 52–53, 54, 55, *96*

Zaba, J., *287*

Subject Index

Absolute permeability, 30, 33, 225
Abu Dhabi, offshore regulations, 8–9
Air, viscosity, 197–198
Alaska, offshore regulations, 6–7
Alberta, 5, 125, 139, 142–143, 162, 163
—, conservation policies, 5
—, D-3 Reef Reservoir, 125
—, Golden Spike Field, 162, 163
—, Harmattan-Elkton Field, 139
—, Rainbow Keg River Field, 163
—, Westerose D-3 Pool, 142–143
—, Wizard Lake D-3A Field, 162, 163
Alcohol flood, 145
Analogy, utilization of, 14–15, 45, 47, 51–53, 113
Application range for secondary recovery, 2, 3, 16–17
Arkansas, Haynesville Field, 112, 114
Austin Chalk Formation, Texas, 122
Australia, offshore regulations, 10–12

Bacon Formation, Texas, 115
Band prediction method, 101–104, 110
— — —, assumptions, 102
— — —, comparison to field performance, 102–104
Block 31 Field, Texas, 159
Bois d'Arc Formation, Texas, 116, 142
Bolivar Coastal Field, Venezuela, 15
Bottom water injection, 14
Bouyancy, effect upon sweep efficiency, '125
Breakthrough, water, 69–70, 75
Brown and White Formation, Texas, 104
Bubble point, compressibility at, 213–214, 216–217
— — pressure, 190–193
Buckley Leverett, 46, 48–50, 52–53, 55, 74–84, 92, 99, 110, 131–138, 150
— —, assumptions, 74
— —, calculation procedure, 75–77
— —, double, 80–83, 92
— —, —, assumptions, 81
— —, —, calculation procedure, 81–84

— —, —, chemical flooding, 83
— —, —, hot-water injection, 81–83
— —, layered systems, 77–81, 99
— —, — —, assumptions, 78
— —, — —, procedure, 78–79
— —, multiple lines of producers, 77
— —, Robert's method, 46, 48–50, 52–53, 55, 77–79
— —, Snyder-Ramey method, 46, 48–50, 52–53, 55, 80
— —, Welge method, 46, 48–50, 52–53, 55, 110, 131–138, 150
— —, —, assumptions, 131
— —, —, correction for prior production, 134–135
— —, —, procedure, 131–135
Buda Limestone Formation, Texas, 122
Burgan Field, Kuwait, 15

California, 6–7, 15, 186–187, 190, 192–193
—, Newhall-Potrero Field, 15
—, offshore regulations, 6–7
—, oil, bubble point pressure, 190
—, —, formation volume factor, 186–187, 192–193
Callaway's equation, 51–52
Canada, 10–12, 110, 187, 191
—, Alberta, 5, 125, 139, 142–143, 162, 163
—, Manitoba, 110
—, offshore regulations, 10–12
—, oil, bubble point pressure, 191
—, —, formation volume factor, 187
Canyon Reef Reservoir, Texas, 123, 126
Capacity (kh), 78, 90, 125
Capillary, desaturation curve, 30
— forces, 32, 173–183
— —, contact angle, 173–183, 226, 232
— —, depression, 173, 174
— —, interfacial tension, 91, 92, 119, 145, 173–182, 226, 231
— —, surface tension, 174–175
— —, wettability, 33, 147, 173–182, 225, 231
— pressure, 29, 32, 118, 119, 181, 236, 237, 246

295

296

— — curves, 29, 236, 237, 246
— —, effect of, 118, 119
— — gradients, 118, 181
Carbon dioxide (CO_2), 16, 204, 205
— —-crude oil mixtures, 204
— —, flood, 16
— —, viscosity, 204, 205
Carbonate reservoirs, fragmental, classification, 250–251
Cedar Creek Anticline, Montana, 111
Channeling of injected fluids (see thief zones)
Characteristics, reservoir (see reservoir characteristics)
Charma Formation, Libya, 114
Chemical injection, 83, 92–94
Clay, compressibility, 220–221
Clearfork Zone Dolomite Formation, Texas, 142
Colorado, Denver Basin, 51
—, Rangely Field, 15, 112
Competitive bidding, 4
Compressibility, 202, 207–223
—, above bubble point, 213
—, below bubble point, 213–214, 222
—, carbonates, 222
—, clay, 220–221
—, effect of solution gas on, 208–210
—, fractured-cavernous carbonates, 208–210
—, gas, 183, 184, 208, 212–213, 214–217
—, isothermal, 207
—, phase, 207–208
—, rock, 202, 219–220, 222–223
—, total system, 207, 211, 213–214, 222
—, undersaturated reservoirs, 214
—, water, 208, 218–219
Computer, 20, 38–40, 110, 113, 120, 142, 158–159, 163
—, data processing, 38–40
—, simulation, 20, 38–40, 110, 113, 120, 142, 158–159, 163
—, —, comparison to field performance, 110, 120, 142, 163
Concession, negociated, 5
Condensing-gas drive (see gas injection, enriched)
Conductance ratio, 108, 109
Connate water saturation, 15
Conservation regulations, 5

Contact angle, 173–183, 226, 232
— —, advancing, 178
— —, effect of temperature on, 232
— —, measurement of, 178
— —, receding, 178
Conversion of units, 264–286
Corey curve, 34–38
Coring, 14, 28, 166
—, analysis of data, 28
—, sampling problems, 14, 166
Cottonwood Creek Field, Wyoming, 117
Coverage (see sweep efficiency)
Craig-Stiles prediction method, 101–104, 108–110
— — —, assumptions, 102
— — —, comparison to field performance, 102–104
Craig-Geffen-Morse prediction method, 101–108
— — —, assumptions, 101
— — —, comparison to field performance, 102–104
Crestal gas injection, 14, 17, 112, 114, 130
Critical velocity, 112–113
Cummins Field, Texas, 114
Cyclic injection, 120, 121

D-3 Reef Reservoir, Alberta, 125
Dahra B Formation, Libya, 114
Dahra Bu Formation, Libya, 114
Darcy equation, 244
Darst Creek Field, Texas, 122
Data collection, 14, 20–28, 141, 166
— —, production records, 20, 24
Definition, conductance ratio, 70
—, contact angle, 173
—, cyclic injection, 120
—, effective permeability, 29, 225
—, gas cycling, 138
—, hydrophilic, 177
—, hydrophobic, 177
—, imbibition, 118
—, miscibility, 145
—, mobility ratio, 70
—, pressure maintenance, 1
—, pressure pulsing, 120
—, primary recovery, 1
—, relative permeability, 29, 225
—, secondary recovery, 1
—, specific surface area, 243

Subject Index

—, wettability, 173
Depth, effect upon secondary recovery, 3, 17
Developer goals, 3, 4, 6–12
Differential gas liberation, 185
Dimensionless permeability, 89
Directional permeability, 2, 13, 69, 113
Dispersed gas injection, 130
Displacement efficiency, 145
Dolomitization, effect upon porosity, 2
Dukhan Field, Qatar, 117
Dump flood, 120–121
Dykstra-Parsons prediction method, 46, 48–50, 52–53, 54, 84–87, 99, 110, 153
— — —, assumptions, 84
— — —, calculation procedure, 84–87
— — — comparison to field performance, 100, 110
— — —, Johnson's charts, 84, 85, 88, 89

Field, Block 31, Texas, 159
—, Burgan, Kuwait, 15
—, Cottonwood Creek, Wyoming, 117
—, Cummins, Texas, 114
—, Darst Creek, Texas, 122
—, Dukhan, Qatar, 117
—, Elk Basin, Wyoming, 115
—, Foster, Texas, 103, 104, 114
—, Fullerton, Texas, 142
—, Golden Spike, Alberta, 162, 163
—, Goldsmith, Texas, 114
—, Haft-Gel, Iran, 117
—, Haynesville, Arkansas and Louisiana, 112, 114
—, Kelly-Snyder, Texas, 17, 115, 126
—, Kirkuk, Iraq, 117
—, Lockout Butte, Montana, 110
—, McElroy, Texas, 100, 111
—, Masjid-i-Sulaiman, Iran, 117
—, Millican, Texas, 162, 163
—, New Hope, Texas, 115
—, Newhall-Potrero, California, 15
—, North Foster, Texas, 114
—, North Virden Scallion, Manitoba, 110
—, Northeast Hallsville, Texas, 92
—, Opelika, Texas, 139
—, Panhandle, Texas, 17, 100, 102, 104
—, Parks, Texas, 162
—, Pegasus Ellenburger, Texas, 123

—, Pennel, Montana, 110
—, Pickton, Texas, 139, 253–263
—, Rainbow Keg River, Alberta, 163
—, Rangely, Colorado, 112
—, Salt Flat, Texas, 122
—, Sholem Alechem Block A, Oklahoma, 112
—, Sharon Ridge Canyon, Texas, 123
—, Slaughter, Texas, 115
—, South Cowden, Texas, 115
—, South Pine, Montana, 110, 111
—, Southwest Lisbon Pettit, Louisiana, 51
—, Spraberry, Texas, 117, 119, 120–121
—, Umm Farud, Libya, 114
—, Waddell, Texas, 115
—, Welch, Texas, 103, 104, 114
—, West Edmond, Texas, 116, 142
—, West Lisbon, Louisiana, 51, 114
—, Wizard Lake D-3A, Alberta, 162, 163
—, Wolfcamp, Texas, 160–161
Fifth Zone, Newhall-Potrero Field, California, 15
Fill-up, 45, 74
Flash gas liberation, 185, 194
Flood efficiency, 5, 13
— —, effect of oil saturation on, 13
Flood life, 56, 74, 77, 79, 86
Flue gas injection, 160
Fluid analysis, 13
— breakthrough, 16
— properties, 20
— velocity, 132
Formation, Austin Chalk, Texas, 122
—, Bacon (Cretaceous), Texas, 115
—, Bois d'Arc, Texas, 116, 142
—, Brown and White (Permian), Texas, 104
—, Buda Limestone, Texas, 122
—, Charma (Ordovician), Libya, 114
—, Clearfork Zone Dolomite, Texas, 142
—, Dahra B (Ordovician), Libya, 114
—, Dahra Bu (Ordovician), Libya, 114
—, Grayburg (Permian), Texas, 114, 115
—, Grayburg-Brown (Permian), Texas, 104
—, Madison (Mississippian), Wyoming, 115
—, Pettit A (Cretaceous), Arkansas, 114
—, Pettit B (Cretaceous), Arkansas, 114

—, San Andres (Permian), Texas, 104, 114, 115
—, San Angelo (Permian), Texas, 115
—, Weber Sand, Colorado, 15
Formation resistivity factor, 245
— volume factor, 130, 185–195, 208, 215, 216
— — —, effect of dissolved gas upon, 185
— — —, natural gas, 183–184, 216
— — —, oil, 185–187, 190–195, 208, 215
— — —, water, 185, 186, 187, 188, 189, 208
Foster Field, Texas, 103, 104, 114
Fractional flow, 74, 75, 76, 101, 132–133
Fractional shrinkage, oil, 187, 194
Fracture, 56, 61, 67–68, 113, 119, 142, 166–172
—, average height, 171–172
—, effect of width, 166–167
—, effect upon production capacity, 166–167
—, effect upon sweep efficiency, 56, 61, 67–68, 113, 142
—, effect upon water/oil ratio, 119
—, tank oil-in-place, 167–169
—, volume, 166, 167–171
Fracture-matrix, 2, 113–122, 141–142, 162
— —, tight matrix, 116, 142
— —, porosity system, 113–122, 141–142, 162
— —, — —, gas injection, 141–142
— —, — —, miscible injection, 162
— —, — —, waterflood, 113–122
Fragmental carbonate reservoirs, classification, 250–251
Frontal displacement, 73–74
Fullerton Field, Texas, 142

Gas, 146–147, 183–184, 197–198, 200–202, 206, 208, 212–214, 216, 217
—, compressibility, 208, 212–213, 214, 217
—, compressibility factors, 184
—, formation volume factor, 183–184, 216
—, pseudo-ternary phase diagram, 146–147
—, viscosity, 197–198, 200–202, 206
Gas-cap drive, 13
Gas cycling (see gas injection)
Gas deviation factor, 183, 216
Gas injection, 2, 3, 4, 14, 16, 17, 26, 112, 114, 130–143, 145, 148–155, 160, 163

— —, applicability of formation to, 2, 3, 16, 17
— —, crestal, 14, 17, 112, 114, 130
— —, cycling, 138–141
— —, dispersed, 130
— —, economics, 4
— —, effect of conservation laws upon, 4
— —, effect of mobility ratio on, 130
— —, enriched drive, 145, 150–155, 163
— —, dispersed, 130
— —, flue gas, 160
— —, high injection pressure, 145, 148–150
— —, pilot test, 130–131, 141, 142
— —, prediction techniques, 131–138
— —, pressure, 130
— —, rate, 130
— —, supply of gas for, 26
Gas/oil ratio, 133, 136–137
— — —, calculation of producing, 133
— — —, cumulative produced, 137
— — —, instantaneous produced, 136–137
Gas velocity, effect upon recovery, 134
Geologic barriers, effect upon sweep efficiency, 66
Golden Spike Field, Alberta, 162, 163
Goldsmith Field, Texas, 114
Grayburg Formation, Texas, 114, 115
Grayburg-Brown Formation, Texas, 104
Grayburg Limestone Field, Texas, 122
Gravity drainage, 134, 147–148

Haft-Gel Field, Iran, 117
Hagen-Poiseuille equation, 243–244
Harmattan-Elkton Field, Alberta, 139
Haynesville Field, Arkansas and Louisiana, 112, 114
Hot-water injection, 81–83
Hydrocarbon system phase relationships, 146–147, 151–161
Hydrophilic, 177, 178
Hydrophobic, 177, 231

Imbibition, rate of, 119
— flood, 117–122
— —, cyclic injection, 120, 121
— —, dump type, 120–121
— —, pressure pulse, 117, 121–122
— —, low rate, 120–121
Improved waterflood, 16, 80, 81–83, 90–94

Subject Index

— —, hot waterflood, 81–83
— —, polymer, 16, 91–92
— —, surfactant, 92–94
Income tax, 7, 9, 12
Injection, gas, crestal, 14, 17, 112, 114, 130
—, —, cycling, 138–141
—, —, dispersed, 130
—, —, pressure, 130
—, —, rate, 130
—, location, 14, 17, 56, 112, 114, 123, 130
—, —, bottom injection, 14
—, —, crestal, 14, 17, 112, 114, 130
—, —, pattern, (see pattern)
—, —, peripheral, 14, 56, 123
—, profile control, 41–42
—, rate, 45, 77, 111, 118
—, water, 14, 56, 69–70, 77, 78, 86, 105, 112, 113, 118, 123
—, —, bottom injection, 14
—, —, breakthrough, 69–70, 75, 123
—, —, cumulative, 77, 78, 86, 105
—, —, gas-oil contact, 112, 113
—, —, peripheral, 14, 56, 123
—, —, rate, 111, 118
Injectivity, 56, 57–60, 61, 62, 63, 64
—, effect of water cut on, 61, 62, 63, 64
—, empirical calculation, 56
Intercrystalline-intergranular porosity systems, 2, 14, 99–115, 141, 157, 159–162
— — —, gas injection, 141
— — —, miscible injection, 157, 159–162
— — —, waterflood, 99–113, 114–115
Interfacial tension, 16, 91, 92, 119, 145, 173–182, 226, 231
— —, control of, 91, 92
— —, effect of 119, 145
— —, effect of temperature upon, 226, 231
Interference test, 113
In situ combustion, 2, 3, 17
— — —, forward, 2, 3, 17
— — —, reverse, 2, 3, 17
Ionic transit time theory, 246
Iran, 8, 9, 117
—, Haft-Gel Field, 117
—, Masjid-i-Sulaiman Field, 117
—, offshore regulations, 14, 56, 123
Iraq, Kirkuk Field, 117

Jamin effect, 182
Joints, effect on waterflood, 113

Kelly-Snyder Field, Texas, 17, 115, 126
Kirkuk Field, Iraq, 117
Kozeny-Carman equation, 245–247
Kuwait, offshore regulations, 8–9
—, Burgan Field, 15

Laboratory analysis, 20, 27–38
Lease, 6–12
—, disposal system, 6–12
—, exploration commitment, 6–12
—, government participation, 6–12
—, relinquishment, 6–12
—, terms, 6–12
Libya, Umm Farud Field, 114
Lockout Butte Field, Montana, 110
Log data, 13, 20
Louisiana, 6–7, 51, 112, 114
—, Haynesville Field, 113, 114
—, offshore regulations, 6–7
—, Southwest Lisbon Pettit Field, 51
—, West Lisbon Field, 51, 114

Madison Formation, Wyoming, 115
Masjid-i-Sulaiman Field, Iran, 117
McElroy Field, Texas, 100, 111
Millican Field, Texas, 162, 163
Miscible displacement, 2, 3, 17, 145–163
— —, alcohol, 145
— —, applicability, 2, 3, 17
— —, enriched gas, 145, 150–155, 163
— —, factors that affect, 145
— —, field performance, 160–163
— —, fracture-matrix systems, 162
— —, high-pressure gas, 145, 148–150
— —, intercrystalline-intergranular systems, 157, 159–162
— —, minimum bank size, 155, 156–157, 162, 163
— —, prediction techniques, 150, 153–154, 158–159, 162
— —, — —, Buckley-Leverett, 150
— —, — —, computer simulation, 158–159, 162
— —, — —, volumetric, 153–154
— —, — —, Welge, 150
— —, slug displacement, 17, 145, 155–158, 162–163
— —, slug size, 157–158, 163
— —, sweep efficiency, 145, 147–150
— —, unsuccessful project, 147

— —, vugular-solution porosity systems, 162–163
Mobility ratio, 13, 16, 56, 69, 70–74, 77, 84, 91, 99, 105, 130
— —, calculation of, 70, 77, 84, 105
— —, control of, 91
— —, effect upon sweep efficiency, 13, 56, 69, 70–74, 91, 145
— —, for gas injection, 130
Montana, 110, 111
—, Lockout Butte Field, 110
—, Pennel Field, 110
—, South Pine Field, 110, 111

Nebraska, Denver Basin, 51
Negociated concession, 5
Neutral zone, offshore regulations, 8–9
New Hope Field, Texas, 115
Newhall-Potrero Field, California, 15
Nigeria, offshore regulations, 10–12
North Foster Field, Texas, 114
North Virden Scallion Field, Canada, 110
Northeast Hallsville, Texas, 92

Offshore regulations, 6–12
Oil, bubble point pressure, 190–193
—, compressibility of, 207, 213–214, 216–217
—, effect of gravity on recovery of, 2, 16, 130
—, effect of polarity on permeability, 225, 227–231
—, effect of variation in reservoir on recovery, 15
—, formation volume factor, 185–187, 190–195, 208, 215
—, gravity, 139, 194, 203
—, phase relationships, 146–147, 151–161
—, solubility of gas in, 209
—, viscosity, 130, 203, 206
Oil-water interfacial tension, 16
Oklahoma, Sholem Alechem Block A Field, 112
Oolites, 2, 92 (see also intergranular-intercrystalline)
Opelika Field, Texas, 139

Panhandle Field, Texas, 17, 100, 102, 104
Parks Field, Texas, 162
Pattern, 14, 17, 56–72, 105, 106, 111, 114, 115, 123, 126, 141, 148, 149, 161

—, direct line drive, 57, 58, 60, 70, 71, 111, 120
—, five-spot, 57, 59–64, 67, 69, 70, 71, 72, 100, 105, 106, 111, 114, 115, 148, 149
—, irregular, 61, 115, 161
—, nine-spot, 58, 60, 111
—, off-pattern, 56, 62–65, 115
—, peripheral, 14, 17, 56, 61, 112, 114, 115, 123, 126
—, selection of, 60–61, 69– 111
—, seven-spot, 57, 60
—, staggered line drive, 57, 61
—, unconfined, 66
Pegasus Ellenburger Field, Texas, 123
Pennel Field, Montana, 110
Pennsylvanian Bend Reservoir, Texas, 162
Peripheral injection pattern, 14, 12, 56, 61, 112, 114, 115, 123, 126
Permeability, 2, 13, 28–38, 69, 75, 89, 91, 113, 117, 119, 120, 166, 225, 243–251
—, absolute, 30, 33, 225
—, anisotropic, 113, 120
—, dimensionless, 89
—, directional, 2, 13, 69, 113
—, effect of fractures on, 113, 117
—, effect of interfacial tension on, 119
—, effect of rock composition on, 250–251
—, effect of surface area on, 243–249, 251
—, effective, 29
—, field determination, 166
—, relative, 28, 29–38, 75, 225, 226–229, 223, 241 (see also relative permeability)
—, —, calculation of, 34–38, 75
—, —, curves, 29, 226–229, 233
—, —, three-phase, 33–34
—, —, two-phase, 32–33
—, variation effect, 84, 85, 86, 89, 91, 92, 99, 100, 110, 111, 122, 123, 130, 142
Permeability-block prediction method (see Stiles)
Permeability ratio, 135
Permeability variation, 84, 85, 86, 88, 89, 100, 110
Permian Basin, Texas, 122
Pettit A Formation, Arkansas, 114
Pettit B Formation, Arkansas, 114
Phase relationships, hydrocarbon systems, 146–147, 151–161
Pickton Field, Texas, 139, 253–263

Subject Index

Pilot test, 20, 28–29, 40–41, 56, 100, 101, 110, 111, 113, 120, 130–131, 141, 142, 147, 162
— —, field examples, 100, 101, 110, 111, 120
— —, gas injection, 130–131, 141, 142
— —, goals, 40
— —, miscible flood, 147, 162
Planning, 20–38
Plugging agents, 42
Poiseuille's law, 178–179
Polymers, 16, 91–92
Pore geometry, 30
— —, tortuosity, 30–32, 245–246
— size, 29
— —, distribution, 33, 246
— structure, 17
Porosity, 2, 28, 169–171
—, calculation of, 169
—, dolomitization effect upon, 2
—, double, 2
—, fracture, 2, 169–171
—, interrhombohedral, 2
— system, fracture-matrix, 2, 113–122, 141–142, 162
— —, intergranular-intercrystalline, 2, 14, 99–115, 141, 157, 159–162
— —, vugular-solution, 2, 122–126, 142–143
Prediction methods, analogy, 14–15, 45, 47, 51–53, 113
— —, Band, 101–104, 110
— —, Buckley-Leverett, 46, 48–50, 52–53, 55, 74–84, 92, 99, 110, 131–138, 150
— —, Callaway's, 51–52
— —, comparison, 45, 46, 47, 48–50, 52–53, 54, 56
— —, computer simulation, 20, 38–40, 110, 113, 120, 142, 158–159, 163
— —, Craig-Geffen-Morse, 101–108
— —, Craig-Stiles, 101–104, 108–110
— —, Dykstra-Parsons, 46, 48–50, 52–53, 54, 84–87, 99, 110, 153
— —, permeability-block, 123–125
— —, rules of thumb, 45, 47
— —, Stiles, 87–90
— —, Tarner, 135–138
Pressure, 1, 3, 17, 113, 130, 139, 166, 183, 197, 198, 200, 214, 216
—, bubble point, 190–193

— buildup, 113, 166
—, capillary (see capillary)
—, gas cycling, 139
— fall-off, 113
— maintenance, 1, 130
—, pseudo-critical, 183, 197, 202, 214
—, pseudo-reduced, 183, 198, 200, 216
Pressure pulsing, 117, 121–122
Primary recovery, 1, 13, 21–22, 24–25
— —, development practice, 21–22
— —, gas-cap drive, 13
— —, operating practices, 24–25
— —, solution-gas drive, 13
— —, water drive, 13
Production date, 24
Productive capacity (see sweep efficiency)
Productivity index, 169
Profit/investment ratio, 5
Proration policies, 5, 16, 145
Pseudo-critical, 183, 197, 202, 214
— pressure, 183, 197, 202, 214
— temperature, 183, 197, 202, 214
Pseudo-reduced, 183, 184, 198, 200, 216
— pressure, 183, 184, 198, 200, 216
— temperature, 183, 184, 198, 200, 216
Pseudo-ternary phase diagram, 146–147

Qatar, Dukhan Field, 117

Rainbow Keg River Field, Alberta, 163
Rangely Field, Colorado, 15, 112
Rate-of-return, 5
Recovery efficiency, 13, 14
— factor, 15
—, ultimate, 5
Recrystallization, 14
Reef, 2–3, 17, 122–123, 147
Relative area factor, 31
Relative permeability, 29–38, 225–241
— —, calculation of, 34–38, 75
— —, curves, 29, 226–229, 233
— —, effect of carbonate material on, 225, 231–232
— —, effect of oil polarity on, 225, 227–231
— —, effect of temperature on, 225, 232–241
— —, effect of water hardness on, 225, 227–231
— —, three-phase, 33–34
— —, two-phase, 32–33

Relative velocity, flood front, 83
Rental fee, offshore, 7, 9, 11
Representative sampling (see coring)
Reserve calculation, 15
Reservoir characteristics, 13, 14, 15, 17, 20–28, 141, 145, 147, 166
— —, coring, 14, 28, 166
— —, data collection, 14, 20–28, 141, 166
— —, determination of, 13, 23
— —, errors in, 15
— —, fluid analysis, 13
— —, geologic, 13, 14, 17, 20, 145, 147
— —, —, heterogeneity, 13, 14, 56, 60, 69, 145
— —, —, lithologic, 13, 20, 147
— —, —, processes, 13
— —, —, stratigraphic, 13, 14, 20, 147
— —, —, structural, 13, 14, 17, 20
— —, log data, 13, 20
Resource development, 3, 4
Response time, 56, 60
Rock properties, 2, 13, 14, 15, 20, 60, 67, 69, 145, 147, 148, 202, 219–220, 222–223
— —, compressibility, 202, 219–220, 222–223
— —, fractures, 14, 56
— —, heterogeneity, 13, 15, 56, 60, 69, 145
— —, recrystallization, 14
— —, secondary solution, 2, 13
— —, stratification, 148
Royalty, 4, 6–12
—, competitive bidding, 4
—, lease practice, 4, 6–12
—, offshore, 6, 8, 12
—, owner interest, 4
—, sliding scale, 4

Salinity, effect upon water properties, 186, 187, 188, 189
Salt Flat Field, Texas, 122
San Andres Formation, Texas, 104, 114, 115
San Angelo Formation, Texas, 115
Saturation, effective, 29
—, gas, 60, 73, 116
—, —, free, 116
—, —, sweep efficiency, 60, 73, 145,
—, irreducible water, temperature effects, 238–239, 241
—, oil, 5, 13, 14, 28, 32, 60, 91, 92, 240
—, —, residual, 92
—, —, sweep efficiency, 5, 13, 14, 28, 32, 60, 91
—, —, temperature effects, 240
—, water, 15, 28, 31–32, 60
—, —, sweep efficiency, 15, 32, 60
Saturation pressure, CO_2-crude oil mixtures, 204
Saudi Arabia, offshore regulations, 8–9
Secondary recovery, 2, 3, 16–17
— —, application ranges, 2, 3, 16–17
— —, — —, effect of depth, 3
— —, — —, effect of oil gravity, 2
— —, — —, effect of pressure, 3
— —, definition, 1
Selective plugging, 41–42
Sholem Alechem Block A Field, Oklahoma, 112
Sharon Ridge Canyon Field, Texas, 123
Simulation, computer, 20, 38–40
Skin damage, 101
Slaughter Field, Texas, 161
Sliding scale royalty, 4
Solution cavities, 2, 13, 113, 166
Solution gas-drive mechanism, 13
South Cowden Field, Texas, 115
Southwest Lisbon Pettit Field, Louisiana, 51
South Pine Field, Montana, 110, 111
Spraberry Field, Texas, 117, 119, 120–121
Steam injection, 17
Stiles prediction method, 46, 48–50, 52–53, 54, 87–90, 99, 100–103, 123–125, 153
— — —, assumptions, 87–88, 101–102
— — —, calculation, 87–90
— — —, comparison to field performance, 100, 103, 123
— — —, layered, 100–101
— — —, permeability-block method, 123, 125
Structural considerations, 17, 22
Systems study (see data collection)
Surface area, specific, 243–249, 251
Surface tension, 174–175
Surfactants, 16, 91, 92–94
Sweep efficiency, areal, 105, 106, 109, 110
— —, breakthrough, 148, 149
— —, coverage, 85, 86, 88, 89, 90
— —, effect of bouyancy on, 125

Subject Index

— —, effect of fluid saturation on, 5, 13, 14, 28, 32, 60, 73, 91, 145
— —, effect of fractures on, 56, 61, 67–68, 113, 142
— —, effect of heterogeneity on, 69–72, 145, 147
— —, effect of imbibition on, 125
— —, effect of mobility ratio on, 13, 56, 69, 70–74, 91, 145
— —, effect of pore geometry on, 147
— —, effect of production on, 62–65
— —, effect of structural dip on, 145, 147
— —, effect of well location on, 13, 55–65, 147
— —, gas injection, 130, 141
— —, miscible displacement, 145, 147–150
— —, off-pattern wells, 61–65
— —, pattern selection, 55, 57–60, 120, 147
— —, unconfined patterns, 66
— —, water injection, 55–74, 91

Tarner prediction method, 135–138
— — —, assumptions, 135
— — —, procedure, 137–138
— — —, required data, 135
Temperature, 183, 197, 198, 200, 214, 216, 225, 232–241
—, effect on relative permeability, 225, 232–241
—, effect upon saturation, 240
—, pseudo-critical, 183, 197, 202, 214
—, pseudo-reduced, 183, 198, 200, 216
Tensleep Sandstone Reservoir, Wyoming, 123, 124
Texas, Block 31 Field, 159
—, Canyon Reef Reservoir, 123, 126
—, Cummins Field, 122
—, Darst Creek Field, 122
—, Foster Field, 103, 104, 114
—, Fullerton Field, 142
—, Goldsmith Field, 114
—, Grayburg Limestone Reservoir, 122
—, Kelly-Snyder Field, 17, 115, 126
—, McElroy Field, 100, 111
—, Millican Field, 162, 163
—, New Hope Field, 115
—, North Foster Field, 114
—, Northeast Hallsville, 92
—, offshore regulations, 6–7

—, Opelika Field, 139
—, Panhandle Field, 17, 100, 102, 104
—, Parks Field, 162
—, Pegasus Ellenburger Field, 123
—, Pennsylvanian Bend Reservoir, 162
—, Pickton Field, 139, 253–263
—, Salt Flat Field, 122
—, Sharon Ridge Canyon Field, 123
—, Slaughter Field, 161
—, South Cowden Field, 115
—, Spraberry Field, 117, 119, 120–121
—, Waddell, Field, 115
—, Welch Field, 103, 104, 114
—, West Edmond Field, 116, 142
—, Wolfcamp Field, 160–161
Thermally reduced oil formation volume factor, 187, 195
Thief zone, 41, 123, 142
Tie line, 146–147
Tortuosity, 30–32, 245–246
Tracer, 111, 113, 142
—, gas, 142
—, water, 111, 113
Trapping mechanism, 14
Trinidad, offshore regulations, 10–12

Ultimate recovery, 5
Umm Farud Field, Libya, 114
Unit recovery factor, 123
United Arab Republic offshore regulations, 8–9
United Kingdom, offshore regulations, 10–12
United States, federal offshore regulations, 6–7
Units, conversion of, 264–286
Unsuccessful waterfloods, 15

Vaporization, oil, 139, 140
Venezuela, 10–12, 15, 187, 190, 191
—, Bolivar Coastal Field, 15
—, LL-370 Area, 190
—, offshore regulations, 10–12
—, oil, 187, 191
—, —, bubble point, 191
—, —, formation volume factor, 187, 191
Viscosity, air, 197–198
—, CO_2-crude oil mixtures, 204, 205
—, gas cycling, 139, 140
—, natural gas, 197–198, 200–202, 206

—, oil, 130, 203, 206
— ratio, 198, 200
—, water, 197, 199
Volumetric prediction method, 153–154
Vugular-solution porosity system, 2, 122–126, 142–143
— — — —, gas injection, 142–143
— — — —, miscible injection, 162–163
— — — —, waterflood, 122–126

Waddell Field, Texas, 115
Water, compressibility of, 208, 218–219
—, cumulative injection, 77, 78, 86, 105
—, formation volume factor, 185, 186, 187, 188, 189, 208
—, gas solubility in, 209–212
—, requirements, 45, 56
—, saturation, 15, 28, 31–32, 60
—, viscosity, 197, 199
Water-drive reservoirs, 13
Waterflood, analogy method, 51–55
—, applicability of formation to, 2, 3, 16, 17, 99
—, breakthrough, 16
—, Callaway's equation, 51–52
—, design, 52–55, 99
—, economic difficulties, 16
—, engineering errors, 15–16
—, fill-up, 45, 74
—, fractures, 94, 113
—, gas cap, 112
—, improved, 80, 90–94

—, injection, 14, 56, 69–70, 75, 77, 78, 86, 105, 112, 113, 118, 123
—, life, 56, 74, 77, 79, 86
—, prediction methods, 46–50, 52–55, 99–126
—, response time, 56, 57
—, rules of thumb, 45, 47
—, supply of water, 25–26, 56
—, sweep efficiency, 55–74
—, unsuccessful, 15–16
—, water requirement, 45
Water/oil ratio (WOR), 77, 79, 90, 118, 119, 125
— — —, calculation, 77, 79, 90, 125
— — —, cut off, 77
Weber Sand, Rangely Field, Colorado, 15
Welch Field, Texas, 103, 104, 114
Welge prediction method, 46, 48–50, 52–53, 55, 110, 131–138, 150
Well spacing, 56, 60
West Edmond Field, Texas, 116, 142
West Lisbon Field, Louisiana, 51, 114
Westerose D-3 Pool, Alberta, 142–143
Wettability, 33, 147, 173–182, 225, 231
Wizard Lake D-3A Field, Alberta, 162, 163
Wolfcamp Field, Texas, 160–161
Working interest, 4, 5
Wyoming, 115, 117, 123, 124
—, Cottonwood Creek Field, 117
—, Elk Basin Field, 115
—, Tensleep Sandstone Reservoir, 123, 124

The Authors

GERALD L. LANGNES is a management sciences consultant for the international division of Mobil Oil Corporation. He received a B.S. in mining engineering from the University of Wisconsin in 1961, and, from the University of Southern California, an M.S. in petroleum engineering (1963) an M.B.A. (1970), and an Engineer's degree—Petroleum Engineer (1971). Since 1962, Mr. Langnes has been employed by the Mobil Oil Corporation in reservoir engineering sections, as a computer communications coordinator, and as a project leader for economic evaluation. He is a member of the Society of Petroleum Engineers of AIME.

JOHN OTIS ROBERTSON, JR., holds a B.S. and an M.S. in petroleum engineering, awarded by the University of Southern California, where he is at present a lecturer while completing a Ph.D. degree in engineering. His professional experience also includes positions as project, production, reservoir and computer systems programming engineer with the Standard Oil Company of California. Author of a number of articles on petroleum engineering, Mr. Robertson is a member of AAAS, Sigma Xi, the American Institute of Mining, Metallurgical, and Petroleum Engineers, and the American Petroleum Institute.

DR. GEORGE V. CHILINGAR is a Professor of Petroleum Engineering at the University of Southern California. He was awarded the degrees of B.E. and M.S. in petroleum engineering (1949 and 1950) and Ph.D. in geology and petroleum engineering in 1956 from the same university. Professor Chilingar has acted as consultant for many companies and held the position of senior petroleum engineering advisor for the United Nations. He was president and vice-president of two engineering firms and, in 1966, was awarded membership in Executive and Professional Hall of Fame. Dr. Chilingar, who is an author of more than one hundred research articles and ten books, is a recipient of numerous awards from various organizations and governments. Several of his books, including three books on carbonate rocks, were translated into foreign languages. Professor Chilingar is a member of the SPE of AIME, AAPG, AGU, GSA, ASEE, SEPM, Geochemical Society, American Institute of Chemists, and the Academies of Science of New York and California.